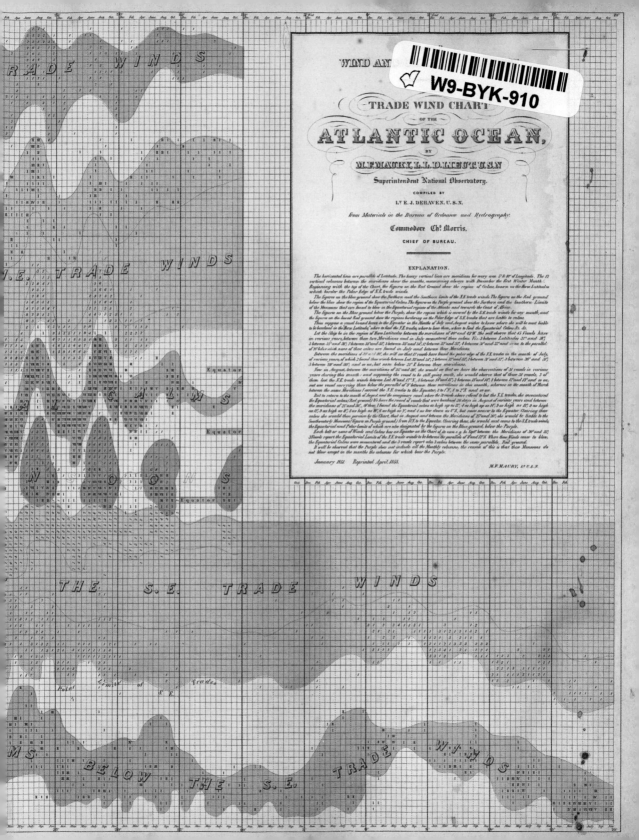

## PRAISE FOR *WEATHER: AN ILLUSTRATED HISTORY*

"FINALLY, someone has done something about the weather. Andrew Revkin and Lisa Mechaley have given us a startlingly fascinating book about how weather got the way it is, and how we've reacted to it, used it, and even helped shape it. There are a hundred captivating stories in this book that are as enlightening as they are fun. Reading them is like seeing the clouds part and the sun come out."

—Alan Alda, longtime host of *Scientific American Frontiers* and a founder of the Alan Alda Center for Communicating Science at Stony Brook University

\* \* \*

"Informative, addictively readable, and never preachy, *Weather: An Illustrated History* tells the fascinating story of humanity's ever-evolving relationship with the earth's climate. Highly recommended."

—Nathaniel Philbrick, National Book Award winner for *In the Heart of the Sea: The Tragedy of the Whaleship Essex*

\* \* \*

"*Weather: An Illustrated History* is a gift of a book—at once fascinating, informative, and surprising."

—Elizabeth Kolbert, Pulitzer Prize–winning author of *The Sixth Extinction*

\* \* \*

"A slim book about a weighty subject with a light touch, *Weather: An Illustrated History* has a wonderfully sprawling cast of characters, from Alexander von Humboldt and 'Snowflake' Bentley to Frankenstein's monster and the editor of the *Farmer's Almanac*. I won't soon forget the image of Benjamin Franklin charging off after a huge dust devil, leaving the rest of his party to gape in astonishment as he repeatedly horsewhipped the whirlwind to see if he could interrupt its progress."

—Charles C. Mann, best-selling author of *1491: New Revelations of the Americas Before Columbus* and *The Wizard and the Prophet*

BOOKS BY ANDREW REVKIN

*The Burning Season: The Murder of Chico Mendes
and the Fight for the Amazon Rain Forest*

*Global Warming: Understanding the Forecast*

*The North Pole Was Here: Puzzles and Perils
at the Top of the World*

# WEATHER

## AN ILLUSTRATED HISTORY

### From CLOUD ATLASES to CLIMATE CHANGE

## ANDREW REVKIN

*with* LISA MECHALEY

STERLING
New York

STERLING
New York

An Imprint of Sterling Publishing Co., Inc.
1166 Avenue of the Americas
New York. NY 10036

ISBN 978-1-4549-2140-0

Distributed in Canada by Sterling Publishing Co., Inc.
c/o Canadian Manda Group, 664 Annette Street
Toronto, Ontario, Canada M6S 2C8
Distributed in the United Kingdom by GMC Distribution Services
Castle Place, 166 High Street, Lewes, East Sussex, England BN7 1XU
Distributed in Australia by NewSouth Books
45 Beach Street, Coogee, NSW 2034, Australia

For information about custom editions, special sales,
and premium and corporate purchases, please contact
Sterling Special Sales at 800-805-5489 or
specialsales@sterlingpublishing.com.

Manufactured in China

2 4 6 8 10 9 7 5 3 1

sterlingpublishing.com

For image credits, see page 209

*This book is dedicated to our sons, Daniel and Jack*

# CONTENTS

# INTRODUCTION

ERE, IN ONE HUNDRED MOMENTS, IS A chronicle of humanity's evolving relationship with, and understanding of, Earth's climate system and the extraordinary weather events that swirl chaotically, and sometimes destructively, within it. For nearly all of human history, the relationship worked in one direction. Climate patterns shifted. Ice sheets, deserts, and coastlines advanced or retreated; extremes of drought, precipitation, wind, or temperature struck and communities thrived, adapted, moved, or faded away. Now, a growing body of science has demonstrated that we are increasingly in a two-way relationship with climate. That momentous transition started as the worldwide spread of agriculture and other human activities changed landscapes suffi-ciently to alter weather patterns millennia ago. The pace and extent of climate changes in decades to come remains unclear. But the atmosphere and oceans have already measurably responded to the heating influence of accumulating greenhouse gas emissions accompanying what Earth scientists have called the "Great Acceleration" in human numbers and resource appetites since around 1950. These gases, most notably carbon dioxide, are transparent to sunlight but absorb some outgoing radiant heat energy.

A full climate chronology would fill volumes. This is more of an exploration, touching on sobering, surprising, even humorous moments in a long and continuing journey of discovery. The goal is to display the range and types of events, insights, and inventions that have punctuated the coevolution of climate and our lives. It is also implicitly a snapshot in time—our generation's moment in this running story. Some of the knowledge gained so far will be upended or rebooted in years and decades to come—just as notions of weather as an expression of the gods' wrath or glee long ago gave way to the understanding of a remarkable system that has both clear patterns (climate) and implicit randomness (the vagaries called weather). As J. Marshall Shepherd, a past president of the American Meteorological Society, likes to put it, "Climate is your personality; weather is your mood."

You'll learn about remarkable insights of brilliant figures with familiar names, like Galileo Galilei and Benjamin Franklin, and discoveries made by obscure but fascinating people, like Mary Anderson, the real estate developer who invented the first windshield wiper, and Wasaburo Ōishi, a Japanese meteorologist who discovered the high-altitude, high-speed jet stream in the 1920s—only to

**An aquatic rescue unit** from the South Carolina Army National Guard was among an array of search teams rescuing people stranded by Texas flooding from Hurricane Harvey in 2017.

have Japan turn it into a weapon, lofting thousands of explosive fire balloons toward the United States during World War II.

In examining the full sweep of what's been learned, and unlearned, about weather and climate through human history, there's one constant: knowledge is forever evolving. It took more than a century of methodical research, measurement, and evolving technology for scientists to move from a basic understanding that some atmospheric gases trap heat to the realization that centuries of warming and sea-level rise could lie ahead should carbon dioxide ($CO_2$) emissions from the burning of fuels and forests not be reduced.

In generations to come, human lives may be so insulated from the weather by technology that it will be considered odd to think people once routinely checked forecasts before venturing out. But, for now, the elements are the one aspect of our environment that nearly every person considers, or is affected by, every single day.

We decided early on to build the chronology around the full sweep of human understanding of the history and workings of the climate system. In going back billions of years, through eras for which evidence is indirect or smeared over millennia by geological wear and tear, we sometimes had to abandon the precision of a conventional chronology—as is most evident in the "moment" between 2.4 billion and 423 million years ago. Those early milestones, while set in a year marked BCE, for before the Common Era, are of course about stretches of time vastly longer than can be comprehended by the human mind—and not

marked with the precision of carbon isotopes or other direct evidence. And of course the final entry, about the end of ice ages, is a speculation on history yet to be written. In most of the chosen milestones, we tried to convey some of the broader significance of discrete events. For instance, the segment about the investigations that undid a longstanding high-temperature record set in Libya in 1922 is as much about the limits of precision in meteorological history as it is about heat.

The narrative thread in the one hundred entries centers on critical scientific insights or disruptive meteorological events, but we included a few more whimsical items—as with the role of weather in music and the lore surrounding a certain groundhog—to capture the full richness of humanity's relationship with the elements.

The historical nuggets that have been left out vastly outnumber our selections. But our hope is that this narrative serves as an appetizer, leading readers to an array of masterful and more comprehensive histories of weather and climate science and lore by writers including Christopher Burt, Brian M. Fagan, James Rodger Fleming, Elizabeth Kolbert, and Spencer R. Weart. And of course there's an enormous range of invaluable online content now, from institutional sources such as the American Meteorological Society, the National Weather Service, and NASA, to weather and climate blogs, including Weather Underground and Realclimate.org.

In producing this book, we've drawn, at points, on the wisdom and words of some friends and colleagues with deep expertise on particular moments in climate history.

Those contributions are indicated by initials at the end of the entries, with details in the endnotes. We invited Howard Lee, a geologist and writer focused on the early Earth, to kick things off. So read on, as we start with the origin of the atmosphere itself—the essential medium in which the dynamics called weather unfold.

# ACKNOWLEDGMENTS

THIS BOOK DRAWS ON THE EXTRAORDINARY scholarship and expertise of dozens of historians, scholars, scientists, and institutions tracking our species' understanding of, and relationship with, the climate system. The references section provides selected sources for each moment and milestone, but we wanted to thank a few individuals, in particular, for their input and insights. Spencer R. Weart, Ph.D., now retired from the directorship of the Center for History of Physics of the American Institute of Physics, has long helped us navigate the sweeping body of research revealing a human role in shaping the climate. His masterful book, *The Discovery of Global Warming*, is an essential resource, particularly the online version at history.aip.org/climate. Professor James Rodger Fleming of Colby College drew on his deep knowledge of meteorological history in kindly and swiftly offering feedback on many items. The early sections of Earth's climate story benefited from scrutiny by Dr. David Grinspoon, senior scientist at the Planetary Science Institute, and Dr. Adam Frank, a professor of astrophysics at the University of Rochester. Any errors or omissions are, of course, our responsibility.

The book also draws heavily on the online archives of the American Meteorological Society, which will celebrate its centennial in 2019, as well as the National Weather Service and other federal agencies focused on weather and climate, along with Britain's Met Office. Excellent starting points for more background are weather.gov/timeline and metoffice.gov.uk, along with a host of National Academy of Sciences reports at this site: nationalacademies.org/climate.

We owe much to the team at Sterling, particularly Meredith Hale, whose keen editing eye helped us avoid several mistakes and keep a consistent tone across such a diverse array of milestones. Stacey Stambaugh did a fine job winnowing art and illustrations that we were able to find, and discovering wonderfully apt imagery when we came up short. The book would not exist if Melanie Madden, a former Sterling science editor, had not persistently circled back to explore ideas with one of us (Andy R.) after first getting in touch way back—in 2012!

Finally, we owe the most to each other, and can only think of one or two moments of slight stress in this, our first writing collaboration—melding more than thirty years of experience in environmental journalism (Andy R.) and environmental and science education (Lisa M.).

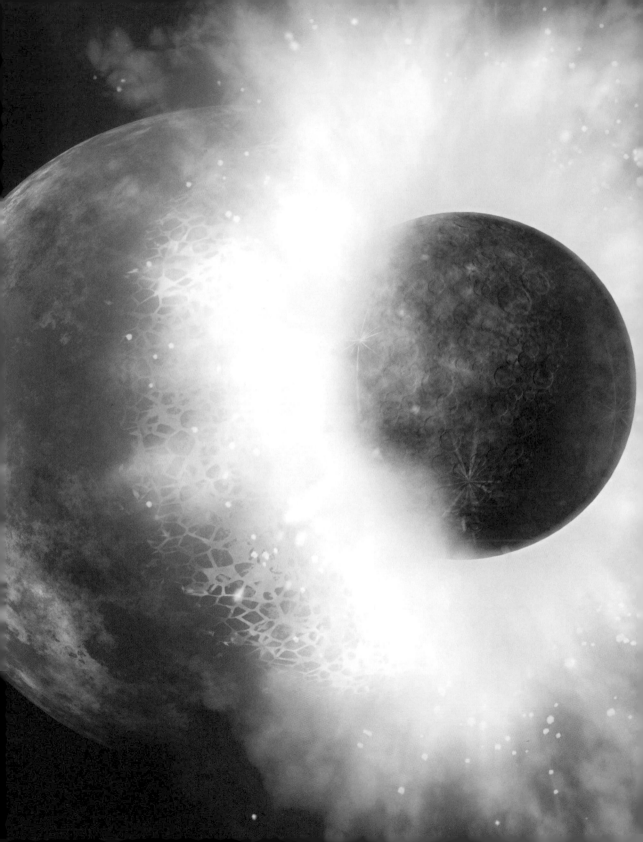

# EARTH GETS AN ATMOSPHERE

To HAVE WEATHER, A PLANET MUST HAVE an atmosphere, so it makes sense to begin this chronology with the origin of ours. The little we know about it is woven from the few strands of evidence science can tie to the earliest years of the solar system: chemical differences between Earth and its presumed building blocks in the form of meteorites, observations of faraway solar systems, and computer simulations that re-create a plausible history of our solar system based on the laws of physics.

These tell us that Earth began forming some 4.567 billion years ago in a slowly spinning cloud of radioactive dust and gas nearly a light year (some 6 trillion miles, or 10 trillion kilometers) across. When the cloud collapsed under its own weight, it formed the sun and, surrounding it, a rotating disk called the solar nebula. Over a few tens of millions of years, dust particles in the disk clumped together, and gravity accreted those clumps into planets, asteroids, comets, and the sun. It was long thought that Earth's first atmosphere formed from gases pulled in by gravity from the surrounding solar nebula. But lately scientists have concluded that most planets' primordial atmospheres are emitted from within— produced in the crucible of pressure and heat generated by incoming colliding material.

Over the first 140 million years or so, bombardment by asteroids would often blow portions of the forming atmosphere away. But the energy from those collisions caused outgassing from molten rock, adding carbon dioxide and carbon monoxide, steam, and sulfur dioxide.

A major reboot occurred around 4.5 billion years ago. An enormous impact with a smaller planet (or perhaps several, according to some studies) transformed Earth's atmosphere into searing-hot rock vapor, which formed a disk of vapor around Earth. By the time the vapor cooled, and either rained as molten rock or coalesced in space as our moon, most or perhaps all of Earth's first atmosphere appears to have been blown into space.

—H. L.

SEE ALSO: Pink Skies and Ice (2.9 Billion BCE), An End to Ice Ages? (102,018 CE)

**Earth's earliest atmosphere** was formed and reformed amid collisions like the one shown in this artist's conception depicting a giant impact similar to the one that may have created the Earth's moon some 4.5 billion years ago.

# WATER WORLD

THE PRICE OF GAINING THE MOON WAS THE loss not only of Earth's first atmosphere, but also much of its water and even elements we don't normally consider easily vaporized, such as lead and zinc. Dry, sterile, and enveloped in boiling magma, Earth went through a hellish time worthy of the name geologists gave to its oldest time period: Hadean eon, for Hades, the ancient Greek god of hell.

And yet, by 4.3 billion years ago, there appears to have been abundant liquid water on Earth's surface. This is known from precise dating and measurements of oxygen isotopes in the oldest minerals on Earth: tiny, purplish, and incredibly durable zircon crystals collected at Jack Hills in Australia.

Geologists remind us that there's a lot of time in deep time, and that span from the great moon-forming collision until oceans formed is a very long time. Here's what scientists think transpired: It took just a handful of years for the rock vapor atmosphere to cool, and about 150,000 years for the global magma ocean to solidify. As it did, vast amounts of carbon dioxide, steam, nitrogen, and sulfur outgassed—sufficient to generate a steam atmosphere. And once the magma ocean crusted over, the steam cooled enough for it to rain, incessantly, for about a million years.

Volcanoes all over the globe spewed lava, which reacted with carbon dioxide in the steamy air. As lava covered lava, flow on flow, more and more carbon was buried in the crust so that, over time, the heat-trapping greenhouse effect diminished. The end result, by around 4.3 billion years ago, was a water world—with a climate conducive to life and a vast global ocean holding up to 26 percent more water than today's oceans, studded with volcanic islands (there were no continents yet) and mountainous impact crater rims. This new, cooler world had weather not radically different from today.

—H. L.

**SEE ALSO:** A Southern Ocean Chills Things (34 Million BCE), Tracking the Oceans' Climate Role (2007)

**This illustration by Howard Lee** shows the vast oceans and scattered volcanic islands that characterized the Earth of 4.3 billion years ago.

# PINK SKIES AND ICE

LIFE APPEARED REMARKABLY EARLY IN Earth's development, possibly as long ago as 4.1 billion years, and definitely by 3.7 billion years ago. However, it took almost a billion years for life to change the climate.

The first life forms were all microscopic. Once some microbes began to extract energy from hydrogen and carbon dioxide to make methane and water, they depleted those gases from the atmosphere—reducing greenhouse gas levels enough to trigger Earth's first-known ice age 2.9 billion years ago.

But as levels of methane rose, the sky turned hazy—and on occasion pink. In the upper atmosphere, ultraviolet rays from the sun split methane molecules, freeing hydrogen, which was light enough to leak into space. Since water is composed of hydrogen and oxygen ($H_2O$), losing hydrogen is like losing water. The amount of water in the ocean slowly diminished.

Fortunately, Earth's water loss didn't last. New microbes called cyanobacteria evolved the trick of photosynthesizing sugar from carbon dioxide and water, about 2.7 billion years ago. Oxygen was a byproduct of this new version of photosynthesis (earlier forms of this reaction did not involve this element). Oxygen is an extremely reactive gas, so it slowly oxidized rocks and chemicals in seawater. As the concentration in the air built, oxygen reacted with methane, producing carbon dioxide and water, preventing hydrogen's escape, and saving Earth's oceans from slowly leaking into space.

As oxygen levels increased, the atmosphere flip-flopped between pink, methane-dominated skies and blue, carbon dioxide–dominated skies. At the same time, erosion and chemical weathering of mountains on Earth's first great single land mass, or supercontinent, known as Kenorland, reduced atmospheric carbon dioxide levels. (In chemical weathering, atmospheric carbon forms a weak carbonic acid in rain, slowly dissolving rocks and resulting in the eventual formation of limestone in seas downstream.) The collapse in levels of Earth's two main greenhouse gases precipitated four distinct "Snowball Earth" moments—periods when the planet's surface nearly or entirely froze and a more temperate climate—between 2.5 and 2.2 billion years ago, until volcanoes restored carbon dioxide levels.

*—H. L.*

SEE ALSO: Earth Gets an Atmosphere (4.567 Billion BCE), The Icy Path to Fire (2.4 Billion–423 Million BCE), Tracking the Oceans' Climate Role (2007)

**The oceans of the prehistoric Kenorland landscape** sustained basic life forms, including simple single-celled organisms and mat-like biofilms of microorganisms such as cyanobacteria.

# FIRST FOSSIL TRACES OF RAINDROPS

FLAT DIMPLES DOTTING ANCIENT SOUTH African rocks are unmistakably raindrop impressions, formed in fresh volcanic ash by a passing shower 2.7 billion years ago. Similar raindrop impressions, identical to modern ones, show up in 2.3 billion-year-old Australian tidal flats and in numerous other examples across geological time. Comparisons to contemporary sediments imply water was flowing in rivers, lakes, and seas, as it does now.

But that should have been impossible.

Back then the sun was only about 80 percent as bright as it is today, so it should have been incapable of warming Earth above freezing. The planet should have been permanently frozen solid from pole to pole. To be warm enough for rain, scientists reasoned, Earth must have had a much thicker atmosphere of heat-holding greenhouse gases than today. But experiments re-creating those same raindrop indentations, and measurements of bubbles in ancient lava, both hint that early atmospheric pressure was probably lower than it is today.

This unexpected early warmth, despite a moderate atmospheric pressure, is known as the Faint Young Sun Paradox. It helps that a thin atmosphere and a global ocean allowed Earth to absorb more sunlight, but the air must also have been rich in greenhouse gases to have retained that warmth. More volcanic carbon dioxide probably erupted from a younger, hotter mantle, while small continents captured far less $CO_2$ through weathering than today. Massive solar flares, eruptions of intense radiation from the surface of the young sun, may have made the greenhouse gas nitrous oxide, or maybe there were novel greenhouse gases produced as nitrogen and hydrogen collided. There may also have been some extra warming from fewer clouds and stronger tides.

Some scientists have even suggested that the faint young sun's power may have been boosted by its being about 5 percent bigger. (Solar wind, coronal mass ejections, and nuclear fusion have made it shrink since then.) That would help explain the coincidence of having liquid water simultaneously on both Earth and more-distant Mars. The trouble is, if all those possibilities were true, the Earth would probably have overheated. Some big questions endure.

—*H. L.*

SEE ALSO: The Rise of Tibet and the Asian Monsoon (10 Million BCE), Shen Kuo Writes of Climate Change (1088 CE)

**A meerkat** sits atop a rock pocked with 2.7-billion-year-old raindrop impressions in South Africa.

# THE ICY PATH TO FIRE

DESPITE THE FAINT YOUNG SUN, EARTH'S climate stayed above freezing for most of the "boring billion" years after oxygen's rise in the atmosphere. Yet oxygen remained at a fraction of modern levels in air, and was largely absent from ocean water. Levels eventually rose to half of modern concentrations around 800 million years ago, tracking the evolution of photosynthetic algae and organisms feeding on it, like amoebae and filter feeders.

Because these complex life forms were much bigger than their microbial rivals, their remains sank deeper in the ocean before bacteria could consume them, carrying carbon and nutrients like phosphorus into deeper waters. This pushed the biological demand for oxygen into deep water, which made oxygen more available on the shallow seabed for the first time, allowing organisms like sponges to take up residence.

Sponges filtered the cyanobacteria-dominated water, allowing sunlight to penetrate, which promoted the rise of oxygen-generating algae. Later, the arrival of jellyfish and zooplankton pushed the chemical cycles involving carbon and oxygen even deeper. Each evolutionary step promoted deeper sinking of carbon-rich remains, more efficiently removing carbon dioxide from the atmosphere and locking it away in the deep ocean and sediments—a process known as the biological carbon pump.

Life's drawdown of greenhouse gases, plus accelerated erosion and rock weathering of equatorial islands and maybe even the colonization of land by lichens, triggered two catastrophic Snowball Earth cycles, beginning 717 million years ago, and a subsequent brief ice age. Most of the planet was icebound for tens of millions of years, leaving only areas near the equator relatively free of ice. Reflecting just how different this world was from today, Siberia and Antarctica were near the equator at the time, and so were among the warmest places on the planet.

Atmospheric oxygen levels remained well below modern levels until plants colonized the land around 470 million years ago. By 423 million years ago, oxygen levels had risen enough to support the first fires, leaving charcoal traces in British rocks.

—H. L.

SEE ALSO: Agriculture Warms the Climate (5,000 BCE), Orbits and Ice Ages (1912)

**Computer artwork** shows the Earth frozen in snow and ice some 590 million years ago, when the continents were in different positions due to tectonic plate movements.

# LETHAL HEAT AND THE "GREAT DYING"

WITH THE RISE OF OXYGEN, LIFE FORMS grew bigger and more energetic. There were leaps of evolution and several mass extinctions, which mostly occurred during unusually active volcanic eruptions known as Large Igneous Provinces. These emitted gigantic quantities of greenhouse gases, warming the climate, acidifying oceans, and often creating widespread oxygen-starved ocean dead zones.

The Permian Mass Extinction, otherwise known as the "Great Dying," was the closest this planet has come to losing complex life altogether.

Before that catastrophe, a menagerie of reptiles roamed the supercontinent of Pangea, which straddled the globe from the South Pole to the Arctic. Ice covered large areas of southern land, fringed by conifer forests. Moisture barely reached the vast continental interior, so desert dunes drifted across parts of Europe and America. That cool climate reversed sharply, however. The first instance occurred 262 million years ago, triggered by eruptions where China is today. But the real lethal heat marking the Great Dying struck 252 million years ago, when Siberia hemorrhaged lava for millennia, burying an area the size of Europe in basalt and ash over two miles (three kilometers) thick.

Acid fog circled the planet and sulfur billowed into the stratosphere, triggering a transient volcanic winter before raining out as highly corrosive acid rain. Greenhouse gas flooded the atmosphere enough to raise global temperatures by 18 degrees Fahrenheit (10 degrees Celsius), making the tropics lethally hot. Carbon dioxide dissolved in the oceans, acidifying them, and they became so starved of oxygen that even seabed-burrowing worms disappeared.

In a preview of the present, fossil-fuel combustion exacerbated the climate change. Magma feeding the eruptions ignited coal and oil deposits, releasing methane and carbon dioxide and spreading fly ash thousands of miles downwind. A staggering 90 percent of marine life, and some 75 percent of land life, went extinct in less than 60,000 years (a geological eye blink). Biodiversity didn't recover until several million years later.

—H. L.

SEE ALSO: Dinosaurs' Demise, Mammals Rise (66 Million BCE), Reefs Feel the Heat (2017)

**Today's level of volcanic activity**, as seen in Hawaii's Kilauea Volcano in 2016, is a pale shadow of past periods when sustained and widespread eruptions shaped climate.

# DINOSAURS' DEMISE, MAMMALS RISE

CHANGE WAS ALREADY IN THE AIR LONG before the dinosaurs disappeared. The climate of the late Cretaceous period—the 79-million-year geologic period that ended the Mesozoic era—had cooled enough for small ice sheets to form in Antarctica. Flowering plants had revolutionized vegetation, mammals were proliferating, and dinosaurs were already in decline.

Then the global climate warmed abruptly (by 14°F/7.8°C in Antarctica) due to massive $CO_2$ emissions from volcanic eruptions in India—similar to those behind the Permian Great Dying, but on a smaller scale. Conditions on land and sea deteriorated for 150,000 years, and species began to go extinct.

By sheer rotten luck, this was the time an asteroid slammed into Earth at a place called Chicxulub in Mexico, 66.021 million years ago.

That was the final blow for the dinosaurs and many other species. For many years, scientists thought they were killed off by a combination of red-hot fallout and a great amount of dust darkening the sky, producing years of global winter. But recent research has cast doubt on the global damage the hot fallout could have done, and has shown there's no evidence for a worldwide conflagration. The impact couldn't have acidified the oceans, and any "impact winter" didn't shut down ocean life—it could only have lasted a couple of years, because fern spores continued to germinate. High-altitude soot from incinerated oil deposits might, however, have induced cooling and drought.

Even if the impact from the asteriod was not globally lethal by itself, however, precise rock dates suggest that the shock wave from that impact also provoked renewed and massive volcanic eruptions in India that caused another abrupt episode of climate warming and ocean acidification.

However the killing was done, it left a landscape dominated by ferns for a thousand years. No dinosaurs (except birds) survived. Mammals were also initially devastated, but within a few hundred thousand years were flourishing. The Age of Mammals, or Cenozoic era, was in high gear and continues today.

—H. L.

SEE ALSO: Earth Gets an Atmosphere (4.567 Billion BCE), Reefs Feel the Heat (2017)

**The impact that created the Chicxulub crater** in Yucatan, Mexico, shown here in an artist's conception, may have caused the extinction of 70 percent of all Earth's species 66 million years ago. The crater is about 112 miles (180 kilometers) wide and was caused by an asteroid or comet core that was 6–12 miles (10–20 kilometers) across.

# THE FEVERISH EOCENE

Around 56 million years ago, the tectonic forces creating the Atlantic Ocean began to split Greenland from Scandinavia. As luck would have it, the rift tapped into the same mantle hot spot that feeds volcanoes in Iceland today.

Magma seeped underground like a gigantic bruise, baking oil-rich sediments offshore from Norway and Ireland, which belched methane into the air through thousands of underwater vents. The North Atlantic must have bubbled like a hot tub.

Methane is a powerful greenhouse gas that converts to another greenhouse gas, carbon dioxide, over a decade. The result? Global temperatures rose by 9 degrees Fahrenheit (5 degrees Celsius), to about 18 degrees warmer than today. In addition, over the course of three to four thousand years, carbon dioxide rose to around three to four times its current levels. That rate was slow enough to prevent a major mass extinction, but some sea creatures and 20 percent of land plants still died off, while mammals evolved smaller body sizes and migrated across continents. This Paleocene-Eocene Thermal Maximum (PETM) event created a climate hot enough for crocodiles and hippolike creatures to live happily just 500 miles (804.6 kilometers) from the North Pole, and for tropical vegetation like palm trees to thrive in the Arctic and in an ice-free Antarctica.

The climate continued to be feverishly hot for several million years, going through repeated bouts of high temperatures known as hyperthermals. These were partly controlled by Earth's regular wobbles as it orbited the sun, and partly by fresh additions of methane baked from sediments resulting from renewed injections of magma. Landscapes fluctuated in pace with these climate cycles, so that Wyoming, for example, alternated between arid salt pans and forest-fringed lakes.

Scientists estimate the amount of carbon dioxide emitted in the PETM was equivalent to humans burning all our fossil fuel reserves. But carbon dioxide emissions from human activities have so far been at a much faster pace than those during the PETM, prompting scientists to project more severe ecological disruption should emissions not decline.

—H. L.

SEE ALSO: An Eruption, Famine, and Monsters (1816)

**Illustration of Trogosus**, an extinct mammal with rodentlike teeth found in Wyoming during the Middle Eocene epoch (48–38 million years BCE).

# A SOUTHERN OCEAN CHILLS THINGS

FOR AS LONG AS THERE HAS BEEN ROCK AND sky, there has been a delicate balance between carbon dioxide production by all the world's volcanoes and the removal of that heat-trapping $CO_2$ from the air through chemical reactions with various kinds of rock—called weathering. When volcanoes dominate, the climate is warmer; when weathering dominates, the climate is cooler. So, when tectonic shifts in the Earth's crust build big continental mountain ranges, erosion and rock weathering tend to slowly cool things off.

When India's slow-motion collision with Asia began around 50 million years ago, the Himalayan Mountains started to rise. Combined with mountains rising in the Americas and Europe, this increased erosion and rock weathering nudged the climate down a long path to cooler temperatures. But if it wasn't for a rearrangement of Earth's oceans, the world may never have experienced the ice ages in which our early human ancestors evolved.

Throughout the dinosaur era, South America and Australia were attached to Antarctica, forcing ocean currents to flow in a tortuous path around the continents. But those attachments finally broke 34 million years ago, allowing waters of a great open Southern Ocean to flow around Antarctica—forming the Antarctic Circumpolar Current. This change rebooted global ocean circulation, boosted marine nutrients, and expanded storage of carbon dioxide in the deep oceans.

Carbon dioxide levels, and global temperatures, plunged.

By 32.8 million years ago, the concentration of carbon dioxide in the air had dropped below 600 parts per million, allowing the Antarctic ice sheet to expand all the way to the sea, and the climate kept cooling. When North and South America reconnected via the Isthmus of Panama 2.8 million years ago, this trend was reinforced, ushering in the Pleistocene ice ages. Concentrations of heat-trapping carbon dioxide dropped below 300 parts per million, and ice advanced in the Northern Hemisphere covering Greenland, much of North America, Scandinavia, and Siberia. These ice sheets fluctuated with over a hundred cold-warm cycles, paced by slight variations in Earth's orientation and orbit around the sun, until the Industrial Era.

—H. L.

SEE ALSO: Medieval Warmth to a Little Ice Age (1100), Tracking the Oceans' Climate Role (2007)

**Starting 34 million years ago**, Australia and South America broke free from Antarctica, with the emergence of a planet-circling open Southern Ocean contributing to cooling and expanded ice.

# THE RISE OF TIBET AND
# THE ASIAN MONSOON

IN MANY PARTS OF THE WORLD, MOISTURE-filled winds flow from the ocean on to the continent during the summer season, producing a lush period with life-giving rains. The name for this phenomenon is *monsoon*, a word derived from the Portuguese *monção*, which is itself derived from the Arabic word for season, *mawsim*. In the tropics, monsoon rainfall is vital to human communities and ecosystems adapted to this cycle. There is no place on the planet where the monsoon cycle is more important than in India and neighboring South Asian countries, where the lives of more than a billion people depend on these rains. Shifts in the timing or pattern of rainfall can result in devastating flooding, or drought and its resulting famine.

The long-term evolution of the monsoon has been controlled by changes in carbon dioxide levels in the atmosphere, with its strength waning 35 to 40 million years ago as carbon dioxide levels slowly fell. But within that time, the slow-motion collision of the Indian subcontinent with Asia raised the Tibetan Plateau some 13,000 feet (4,000 meters), dramatically changing climate patterns as the vast, high-altitude plains absorbed the heat from the summer sun. This led to increased onshore wind flow and stronger monsoon rains over India. According to analysis by Brown University geologist Steven C. Clemens and others, Indian monsoons similar to those experienced today likely evolved sometime between 12 and 10 million years ago.

On much shorter time scales, occasional abrupt decreases in monsoon strength coincided with pulses of melting ice sheets in North America and Greenland at certain points during the last ice age, which ended 11,700 years ago. When vast volumes of fresh water—which is less dense than seawater—flowed into the high-latitude North Atlantic, this slowed the oceanic currents carrying warm tropical water north. The meteorological impacts of these events spread across Europe and Asia and into the monsoon regions via westerly winds. Efforts are currently underway to assess the impact of today's rising carbon dioxide levels and future freshening of the North Atlantic on the Indian and Asian monsoon systems.

SEE ALSO: The Fertile Crescent (9,700 BCE), Tracking the Oceans' Climate Role (2007)

**A woman walks** down a flooded street during monsoon season in New Delhi, India, January 2008.

# CLIMATE PULSE PROPELS POPULATIONS

MANY MYSTERIES STILL SURROUND HOW and when modern humans fanned out around the planet from the species' birthplace in Africa. Until recently, the longstanding notion was that a single wave of migration occurred around 60,000 years ago, with branches spreading here and there over the continents.

Researchers posited that the dispersal occurred when a lush interval created a grassy, verdant corridor across the hostile deserts of North Africa and Arabia. A study published in *Nature* in September 2016 by researchers at the University of Hawaii at Manoa built on this narrative, pushing back the start of the exodus tens of thousands of years, closer to 100,000 BCE. Using computer models of the region's climate and ecosystem, the scientists argued that humans moved out of Africa in four waves: the first starting 106,000 to 94,000 years ago; the second, 89,000 to 73,000 years ago; the third, 59,000 to 47,000 years ago; and the last, 45,000 to 29,000 years ago.

Three independent studies comparing DNA from cultures around the world, published in the same issue of *Nature,* allowed for this picture of a pulsed human diaspora, but reaffirmed that modern human populations all split off from a single pulse sometime between 65,000 and 55,000 years ago.

Intriguingly, new paleoclimate evidence suggests that this particular time interval was not a wet time. In late 2017, Jessica Tierney and Paul Zander of the University of Arizona and Peter deMenocal of Columbia University examined more detailed climate clues, and found a particularly dry and cool spell 60,000 years ago that corresponded with the time of migration suggested by DNA studies. This evidence suggests that, rather than drawn out of Africa by wet conditions, early humans may have been pushed out by drought.

The story will surely continue to evolve as new climatic, archaeological, and genetic evidence come to light.

**SEE ALSO:** North Africa Dries and the Pharaohs Rise (5,300 BCE), Ice Ages Revealed (1840)

**Two giraffes carved in a rock** face in the center of the Sahara Desert around 6,000 BCE. Starting 100,000 years ago, according to recent research, pulses of wetter conditions turned North Africa and the Arabian Peninsula into bountiful ecosystems, creating something of a migratory "valve" for humans leaving Africa, and a path to Eurasia.

# A SUPER DROUGHT

IN 1960, SCIENTISTS FROM DUKE UNIVERSITY pushed sediment-coring tubes into the soft, muddy floor of a shallow bay of Lake Victoria in Africa—the largest tropical lake in the world, with fisheries along its shores in Uganda, Kenya, and Tanzania feeding more than 20 million people. They were surprised when their core sampler was stopped by a dense layer of gray clay that could only have formed in the open air. Later studies traced the dry layer into the deepest part of the lake and confirmed that around 15,000 to 14,000 BCE the vast body of water had vanished during a massive, century-scale drought at the end of the last ice age.

The desiccation of Lake Victoria was surprising for several reasons. The lake contains hundreds of species of fish found nowhere else, meaning they must have evolved at an explosive rate after that great drying. More surprising was the sheer scale of the drought. When Lake Victoria vanished, so too did the Nile River's other major source, Lake Tana in Ethiopia. The world's longest river must have shriveled along with the lakes. The drought also shrank lakes in the rest of tropical Africa and in the Jordan Valley, as well. Layered cave deposits likewise registered a weakening of monsoons throughout southern Asia, and geneticists have discovered signs of a crash in human populations in India around that time period.

Early interpretations of evidence from sites in northern Africa suggested that the rain belt had merely shifted south. However, the more southerly African sites demonstrate otherwise. The entire Afro-Asian monsoon system, which now supports more than half of all humanity, had essentially collapsed in one of the most widespread and catastrophic tropical droughts in the history of anatomically modern humans. Unfortunately, we are not yet sure why it happened.

The megadrought coincided with a sudden disintegration of ice sheets, known as a Heinrich event, but it remains unclear whether the collapse caused the drought or simply accompanied it. Other lake studies have since revealed unnerving drought patterns in other parts of Africa. In 2009, cores extracted from a crater lake in Ghana revealed that century-long extreme droughts have occurred repeatedly over the last several thousand years in that now-populous region of the continent.

—C. S.

SEE ALSO: A Dry Discovery (1903), The Dust Bowl (1935)

**Lake Victoria**, seen in this 2013 *Terra* satellite image, is the largest lake in Africa, supporting the livelihoods of some 35 million people. Around 17,000 years ago, it completely dried up.

# THE FERTILE CRESCENT

THE END OF THE MOST RECENT ICE AGE 11,700 years ago (approximately 9,700 BCE) marks the beginning of the Holocene Epoch in the geological time scale. It also set the stage for humans to shift increasingly to stable settlements relying ever more on agriculture. Changing climate conditions prompted people to migrate from the Arabian Peninsula northward, seeking reliable water supplies. The Tigris and Euphrates Rivers, running parallel to each other for several hundred miles, offered a particularly productive zone. The area from the Nile through Jordan and Israel and along these two rivers eventually became known to historians as the Fertile Crescent. Some of the first cities emerged in this region, bringing written language, science, and organized religion. Prosperous kingdoms sprang up between the Tigris and Euphrates, the region known as Mesopotamia, "land between the rivers."

The region was home to wild plants that were the source of critical Neolithic "founder crops" of early agriculture—emmer wheat, barley, flax, chickpeas, peas, lentils—as well as four of the most important species of domesticated animals: cows, goats, sheep, and pigs. Inventions critical to human advancement originated here, as well, including glass, the wheel, and irrigation.

The term "Fertile Crescent" was popularized in the early twentieth century through textbooks written by James Henry Breasted (1865–1935), an archaeologist at the University of Chicago. In his book *Survey of the Ancient World,* first published in 1919, he captured the pull of the region's climate and geography—and the conflict that often resulted from it: "The history of Western Asia may be described as an age-long struggle between the mountain peoples of the north and the desert wanderers of these grasslands—a struggle which is still going on—for the possession of the Fertile Crescent."

Indeed, a century later, tensions still swirl in these same regions. Conflicts in communities already torn by clashing cultures or ideologies have been intensified as dams, diminishing groundwater supplies, and drought—possibly worsened by climate change—strain water supplies and agricultural production.

SEE ALSO: Climate Pulse Propels Populations (100,000 BCE), Agriculture Warms the Climate (5,000 BCE)

**Painting from the tomb** of Egyptian artisan Sennedjem shows him plowing his field in Deir el-Medina, Egypt, an important area for early agriculture.

# NORTH AFRICA DRIES AND THE PHARAOHS RISE

NORTH AFRICA WAS AT A HUMID PEAK around 8,500 to 5,300 BCE, with pastoral hunter-gatherer and herding communities drawn to a fertile, lake-dotted savanna grazed by giraffes, antelopes, and elephants. Where desert dominates now, hippos wallowed in marsh-lined rivers. Dispersed hunting and fishing communities and a growing array of agricultural settlements dotted the region.

The historian Roland Oliver (1923–2014), in his 1999 book *The African Experience: From Olduvai Gorge to the 21st Century,* described how the highlands of the central Sahara Desert were cloaked with forests dense with oak, walnut, lime, and elm trees, while olive, pine, and juniper trees covered lower slopes and grass-bordered, fish-filled rivers laced the valleys.

Confirming the verdant conditions, in 2014, a research team including Christopher McKay, a scientist from NASA's Ames Research Center who has roamed a host of Mars-like environments on Earth, discovered mineral deposits outlining the shores of lakes that existed between 8,100 and 9,400 years ago in what is now the driest part of the Sahara. The site, located in southwestern Egypt, is near remarkable rock art that some experts think depicted people swimming.

A gradual drying occurred in the region from 5,300 to 3,500 BCE, and people settled increasingly along the Nile. The first farms in the region appeared. Starting around then, the age of the pharaohs developed along the Nile, with their civilization lasting 3,000 years.

Research by Columbia University scientists in 2013 found strong evidence that a southward shift in a belt of low pressure where the trade winds of the Northern and Southern Hemispheres meet, called the Intertropical Convergence Zone, tipped the climate toward the dry, hot conditions that have dominated since. A subtle shift in Earth's orbit appears to be behind the change.

**SEE ALSO:** Agriculture Warms the Climate (5,000 BCE), A Dry Discovery (1903), Settling a Hot Debate (2012)

**Cave paintings found in southwestern Egypt** and dated to roughly 5,000 BCE depict the last humid era in North Africa before the climate slid into its current dry norm. One set of images, in the "Cave of the Swimmers," appears to show humans enjoying a water hole or lake long since vanished.

# AGRICULTURE WARMS THE CLIMATE

BY THE TIME THE ICE RECEDED AT THE end of the last ice age, humans had already occupied every continent, except Antarctica, for millennia. But now humankind found itself in a warm and stable climate conducive to domesticating crops. Eventually, humans came to depend on crops for almost all their food, a trend that began in the Middle East around 6,000 BCE, and spread across Europe, China, and the rest of the world over succeeding centuries.

As populations rose and agriculture spread, people burned forests to clear space for farming. As they did, around 5,000 BCE (according to the work of William Ruddiman [b. 1943] and others), levels of heat-trapping carbon dioxide and then methane began to rise in the atmosphere. This is revealed by tiny bubbles of air trapped in ancient glacial ice in Greenland and Antarctica, in archaeological remains, in ancient pollen, and other clues. By the Roman era, half the forests of Europe had gone, while in China, deforestation forced people to burn coal instead of wood for heat by the year 400. The rise in methane concentrations coincided with the spread of irrigated rice cultivation from the Yangtze River valley across Asia starting in 3,000 BCE.

Following a decade of scientific debate, Ruddiman and eleven co-authors laid out a comprehensive argument (2016, in *Reviews of Geophysics*) that this human-generated greenhouse-gas buildup began to warm the climate just enough to delay inevitable cooling, as Earth's orbit was slowly shifting toward producing the next cold cycle of the ice ages, as it had some hundred times in the preceding 2.6 million years.

A critical piece of evidence was the contrast between the distinctive rise in $CO_2$ concentrations starting in 5,000 BCE and methane concentrations 2,000 years later with *downward* trends in levels of those gases during equivalent periods in previous warm intervals between ice ages.

—H. L.

SEE ALSO: Coal, $CO_2$, and the Climate (1896), Orbits and Ice Ages (1912)

**Terraced rice fields**, a common farming technique in much of Asia. As people cleared and tended farmlands in these ways thousands of years ago, greenhouse gases built, stemming a long-term cooling trend.

# ARISTOTLE'S *METEOROLOGICA*

350 BCE

ARISTOTLE (C. 384–C. 322 BCE), ONE OF humanity's great polymaths, explored a dizzying range of subjects, from ethics and mathematics to botany and agriculture, from politics and medicine to dance and theater. He traveled widely along Europe's Mediterranean shores, developing an intimate awareness of environmental dynamics. His observations of environmental processes were assembled in a landmark work, *Meteorologica*.

Today, the word *meteorology* describes the study of weather, although Aristotle had a much wider scope, writing on "all the affections we may call common to air and water, and the kinds and parts of the earth and the affections of its parts."

But weather and the environmental factors shaping it were a prime focus. He proposed the Earth was split into climatic zones, based on distance from the equator, of "frigid," temperate, and "torrid" conditions. His treatise contains one of the first descriptions of the hydrologic cycle:

Now the sun, moving as it does, sets up processes of change and becoming and decay, and by its agency the finest and sweetest water is every day carried up and is dissolved into vapor and rises to the upper region, where it is condensed again by the cold and so returns to the earth.

Not having the tools of his successors, Aristotle made plenty of mistakes, including presuming the Milky Way and comets were in the atmosphere. But his keen eye for detail was evident in page after page of methodical descriptions and deconstructions of rainbows and halos, thunder and lightning, hail and snow. The era in which the wrath or beneficence of Zeus or Aeolus explained weather was starting to give way to an analytical approach.

SEE ALSO: China Shifts from Mythology to Meteorology (300 **BCE**), Shen Kuo Writes of Climate Change (1088 **CE**), Meteorology Gets Useful (1870)

*The School of Athens* (1509) by Italian painter Raphael shows Greek intellectuals, including Aristotle, author of *Meteorologica*, engaging in lively discussion.

# CHINA SHIFTS FROM MYTHOLOGY TO METEOROLOGY

SOME OF THE OLDEST CHINESE WRITTEN records touching on weather lore are inscribed on "oracle bones" dating from the latter part of the Shang dynasty, which ended in 1050 BCE. These objects, made of ox bone or the flat belly shells of turtles, were sometimes used by priests to divine crop planting insights or hints of seasons to come.

Paralleling the evolution of ideas about the planets' dynamics in ancient Greece, Chinese mythology related to weather began to give way to a more analytical approach. Chinese scholars and priests began using observations to track the seasons, measuring the shadow cast by a pole at noon. When the shadow was at its longest, meaning the sun was at its lowest position in the sky, the winter solstice had arrived. The shortest shadow indicated the summer solstice.

By 300 BCE, Chinese astronomers had developed a calendar based on the sun's position in the zodiac. This calendar divided the year into twenty-four festivals, each associated with a different type of weather. Terms such as *major heat* and *minor cold* described changes in temperature throughout the year. Precipitation and harvest times were also represented in the calendar.

During the Han dynasty, in the first century CE of Western calendars, the philosopher Wang Chong (27–c. 100 CE), building on older texts, tried to dispel old notions that weather reflected heaven's temperament. He wrote a passage in his classic text *Lunheng* that beautifully captures the hydrologic cycle:

> As to this coming of rain from the mountains, some hold that the clouds carry the rain with them, dispersing as it is precipitated (and they are right). Clouds and rain are really the same thing. Water evaporating upwards becomes clouds, which condense into rain, or still further into dew.

He was largely ignored until the nineteenth and twentieth centuries, when China's modern scientific enterprise emerged.

SEE ALSO: Aristotle's *Meteorologica* (350 BCE), Shen Kuo Writes of Climate Change (1088 CE)

**Dragon Amid Clouds and Waves**, ink on silk scroll, by an unidentified Chinese artist during the Ming Dynasty (1368–1644). In ancient China, mythology and lore often influenced ideas about weather.

# SHEN KUO WRITES OF CLIMATE CHANGE

SOME OF THE MOST IMPORTANT INSIGHTS into today's climate and what lies ahead have come from the field of paleoclimatology—studies of durable clues to past conditions preserved in the layered sediment of lakes or seabeds, tree rings, the chemical composition of fossils, bubbles of ancient air trapped in glacial ice, and other natural repositories.

Much of that science has its roots in the nineteenth century in the West. But perhaps the earliest written insight gleaned from nature about the capacity for the climate of a particular region to change over time came from a Chinese scholar, engineer, philosopher, and government official: Shen Kuo (1031–95).

In *Dream Pool Essays*, a wide-ranging treatise written in 1088, he proposed explanations for tornadoes and rainbows, and observed how lightning could melt metal objects in a house while only scorching the walls. He had ecological insights, such as noticing how rising demands for wood were eroding forest resources.

But perhaps most remarkable of all, he made a pioneering observation that the climate in one place was not constant, building on observations made years earlier in a town where a high riverbank had given way, revealing a mystery:

[U]nder the ground a forest of bamboo shoots was thus revealed. It contained several hundred bamboo with their roots and trunks all complete, and all turned to stone. . . . Now bamboos do not grow in Yanzhou. . . . Perhaps in very ancient times, the climate was different so that the place was low, damp, gloomy, and suitable for bamboos. On the Jin-hua Shan in Wuzhou, there are stone pine cones and stones formed from peach kernels, stone bulrush roots, stone fishes, crabs, and so on, but as these are all (modern) native products of that place, people are not very surprised at them. But these petrified bamboos appeared under the ground so deep, though they are not produced in that place today. This is a very strange thing indeed.

His observation represented a profound shift in thinking from longstanding views, both in the East and West, that the basic workings of nature, while sometimes tempestuous or dramatic, were essentially unchanging.

SEE ALSO: Ice Ages Revealed (1840), Peat Bog History (1841)

**A bronze bust of Shen Kuo** (1031–95), a scientist and statesman of the Song Dynasty. After seeing fossilized bamboo in a region without this plant, he posited that climate—long assumed to be a constant—could change.

# MEDIEVAL WARMTH TO A LITTLE ICE AGE

A MEDIEVAL WARM PERIOD OF SEVERAL centuries, centered around 1100 CE, was first proposed in 1965 by a pioneering researcher of past climates, Hubert Lamb, who also noted evidence of a subsequent decrease of temperature until, as he wrote, "between 1500 and 1700 the coldest phase since the last ice age occurred."

Decades of sustained research has filled in some blanks but also raised fresh questions about the scope and geographic extent of the warm period and subsequent cool stretch, dubbed the Little Ice Age.

A 2016 paper by Raymond Bradley of the University of Massachusetts (and two co-authors) proposed a different way of thinking about climate around that time, positing that the span from 725 to 1025 CE could be seen as a Medieval Quiet Period, the only stretch of several centuries in the last two thousand years without some big jog to the system from solar variations or big volcanoes.

There are still competing theories about the mix of forces that caused a momentous transition toward cooling, which coincided in Europe with a host of trials, including the Great Famine from 1315 to 1317 and the Black Death of 1347 to 1351.

Recent studies point to cooling veils of particles released during several spasms of intense volcanic activity as the likely trigger, with an expansion of sea ice in the Arctic amplifying the impact.

It has become clear, too, that some short-lived periods of disruptive weather simply emerged from chaotic variability in the planet's complex climate system. This appears to have been the case from 1430 to 1440, according to a paper published in the journal *Climates of the Past* in 2016. During this time, a period of exceptional cold sparked famine and disease outbreaks across Europe. The authors warned, "Our analysis of the subsistence crisis of the 1430s shows that societies that are not prepared for adverse climatic and environmental conditions are vulnerable and may pay a high toll."

SEE ALSO: London's Last Frost Fair (1814), Ice Ages Revealed (1840), Orbits and Ice Ages (1912)

*Winter Landscape with Bird Trap* by Flemish painter Pieter Brueghel the Younger (c. 1564–c. 1638) shows townsfolk walking on a frozen lake, during a cold phase in European climate history.

# THE AGE OF SAIL

SOME NAMELESS INNOVATOR DRIFTING ON a raft or boat thousands of years ago raised cloth against the wind, putting this variable but powerful force to use for the first time and giving birth to the basic concept of sailing.

Sails cast shadows along the Nile as early as 3,400 years ago, memorialized in tomb artwork. Polynesian civilization spread over dispersed Pacific Islands through mastery of sailing canoes and subtle navigation clues. Many of the world's first major powers, from China to Arabia to the Mediterranean, arose substantially through their capacity to ride the wind at sea.

Increasingly sophisticated hull and rig designs emerged. In an extraordinary, if brief, display of maritime power, from 1414 to 1433, China dispatched a fleet of more than sixty junk-rigged ships through Southeast Asia to Africa (and possibly south and into the Atlantic, according to a few clues). At the fleet's core was an immense nine-masted, 400-foot-long (121.9-meter) treasure ship.

By contrast, the *Santa Maria*, the largest of the three vessels on Columbus's 1492 journey, was just around 115 feet (35 meters) long. But China's leaders turned inward while European nations pressed out. The Western "Age of Sail," from 1571 to 1862, saw vastly larger numbers of ships traverse the world's farthest seas.

One of the greatest tests of forces and tactics came in 1588, when Spain's "Invincible Armada" of 130 warships left Lisbon aiming to take control of the English Channel and deposit 20,000 soldiers on British soil. Fortuitous weather and longer-range cannons gave Queen Elizabeth's forces the upper hand.

By the 1800s, maritime commerce was creating a great burst of economic and cultural globalization, with Chinese tea and California gold fueling the flow of high-speed clipper ships. The Suez Canal, opened in 1869, was a great disruptor, and then, of course, came the age of steam.

SEE ALSO: Beaufort Classifies the Winds (1806), Putting Wind to Work (1887)

**British ships and the Spanish Armada** engage in this pre-1700 painting of the famous 1588 battle, in which a change in the weather favored England.

# THE INVENTION OF TEMPERATURE

As long ago as 170 CE, a Greek physician and scientist, Galen of Pergamum (c. 130–c. 216 CE), proposed defining a standard baseline temperature by mixing equal parts of boiling water and ice, with a scale demarcating four degrees of heat and four degrees of cold relative to that central condition.

But the idea of measuring heat in a quantifiable, consistent way—the invention of temperature—emerged far more fully in Venice, Italy, at the start of the seventeenth century. There, Galileo Galilei (1564–1642) and a cluster of contemporaries combined passions for problem-solving and tinkering and produced a novel set of instruments called thermoscopes that were just a short step away from a modern thermometer.

This was part of a broader movement toward the "mathematization of nature," according to Albert Van Helden, a professor emeritus at Rice University whose analysis of Galileo's work can be found online at The Galileo Project (galileo.rice.edu). Van Helden describes the earliest versions of the thermoscope, which allowed the user to gauge temperature changes through the expansion of water, as little more than novelties.

He quotes a passage written in 1638 by a contemporary of Galileo, Benedetto Castelli (1578–1643), who recalled seeing the thermoscope in Galileo's hands around 1603:

He took a small glass flask, about as large as a small hen's egg, with a neck about two spans long [perhaps 16 inches, or 40.6 centimeters] and as fine as a wheat straw, and warmed the flask well in his hands, then turned its mouth upside down into a vessel placed underneath, in which there was a little water. When he took away the heat of his hands from the flask, the water at once began to rise in the neck, and mounted to more than a span above the level of the water in the vessel. The same Sig. Galileo had then made use of this effect in order to construct an instrument for examining the degrees of heat and cold.

The design of the device was refined in successive years by others, particularly Santorio Santorio (1561–1636) and Galileo's friend Gianfrancesco Sagredo (1571–1620). A numerical scale was added to the slender neck, and it did not take long before the first meteorological temperature observations began.

SEE ALSO: Fahrenheit Standardizes Degrees (1714), Settling a Hot Debate (2012)

**Galileo Galilei** (1564–1642), pictured here in an eighteenth-century oil portrait, was one of the first scholars to propose that temperature could be precisely measured and to test ways to accomplish this feat.

# DECIPHERING THE RAINBOW

RAINBOWS HAVE BEEN SOURCES OF WONDER, curiosity, and mythology throughout human history—the Norse path connecting Earth to Asgard, home of the gods; an approving sign from God after Noah built an altar following the Great Flood; a snakelike creator in Australian aboriginal stories.

In Western literature, Aristotle was the first to offer an explanation of the nature and cause of rainbows. In his text *Meteorologica*, he proposed that a rainbow is an unusual reflection of sunlight off of water droplets inside rain clouds. This notion endured for seventeen centuries before a German monk named Theodoric of Freiberg (1250–1310) came up with a different theory in 1304. Theodoric proposed that each water drop is capable of producing a rainbow. Devising experiments with prisms, screens, and spherical flasks of water, Theodoric was able to determine that the path light takes from the sun through a drop of water to the human eye creates a rainbow.

Theodoric's insights remained relatively unknown until rediscovered by Rene Descartes (1596–1650). Descartes was a French mathematician and scholar, and is considered by many to be the father of modern philosophy. In his seminal 1637 essay, Les Météores, Descartes deconstructed the physics of rainbows, describing how light is refracted as it enters a spherical drop, reflected by the curved back surface of the drop, and then refracted again as it moves from water back into air. Descartes built on the spherical-flask experiments of Theodoric, but precisely calculated the path sunlight takes from different points through the flask to determine the angles of refraction.

Isaac Newton (1643–1727) and others later provided insights to how different wavelengths, and thus colors, of light mix into what is perceived as daylight. In more recent times, added complexities have emerged as scientists have probed further into how rainbows form. For instance, larger raindrops are not spheres, but flattened on the underside as they fall, due to air resistance. High-quality digital cameras have captured images of other unusual rainbow features, demonstrating that more mysteries remain.

SEE ALSO: Aristotle's *Meteorologica* (350 BCE), Proof of Electrical "Sprites" (1989)

**Throughout human history**, rainbows have inspired both wonder and scientific speculation—and art, including this landscape, *Rainy Season in the Tropics*, by the American painter Frederic Edwin Church (1826–1900).

# THE WEIGHT OF THE ATMOSPHERE

FROM THE TIME OF THE ANCIENT GREEKS through Galileo, scholars had presumed that air had no weight. That changed when experimental efforts to divine the nature of a vacuum yielded a revolutionary, broader conclusion by the Italian physicist and mathematician Evangelista Torricelli (1608–47).

Galileo had been stymied by a physical puzzle: Well diggers were finding it impossible to siphon water from more than about 30 vertical feet (9 meters).

To test whether a vacuum was involved, another scientist from Florence, Gasparo Berti, created an experiment with a water filled vertical lead pipe, the bottom end of which was immersed in an open cistern of water. A sustained debate built around the nature of the empty space formed at the top of the pipe.

Torricelli had moved to Florence in 1641 to serve as Galileo's secretary and assistant for what turned out to be the last few months of the master scientist's life. Galileo died in 1642, at age seventy-seven, and Torricelli took up the puzzle, creating a more compact version of the device using a glass tube and mercury instead of water.

Further experiments generated a vacuum. But more importantly, Torricelli came away with a profound insight: The liquid was not being pulled up the tube by some mysterious force within. It was being pushed upward by the weight of the atmosphere pressing down on the liquid from without. In a letter to a colleague on June 11, 1644, Torricelli wrote:

> We live immersed at the bottom of a sea of elemental air, which by experiment undoubtedly has weight, and so much weight that the densest air in the neighborhood of the surface of the earth weighs about one four-hundredth part of the weight of water.

Torricelli observed that the mercury level changed along with atmospheric conditions. Notes from a lecture he later gave show he elucidated a link between pressure and weather. With these words, he laid the groundwork for meteorology: "Winds are produced by differences of air temperature, and hence density, between two regions of the earth."

SEE ALSO: Earth Gets an Atmosphere (4.567 Billion BCE), Meteorology Gets Useful (1870)

**Torricelli** devised how to determine atmospheric pressure by measuring changes in the height of a column of mercury in a tube, inverted in a dish of the liquid metal.

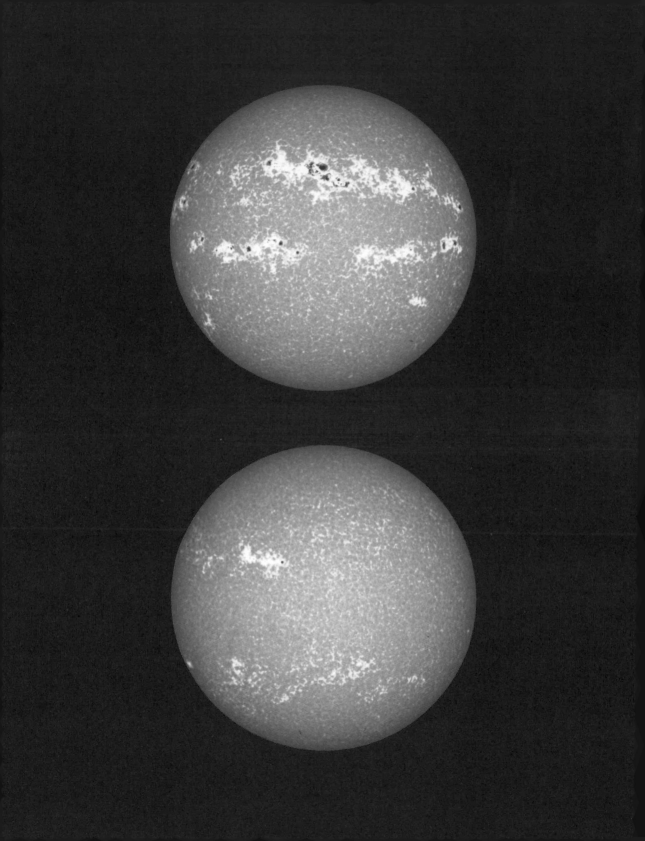

# A SPOTLESS SUN

A S STARS GO, THE SUN IS REMARKABLY steady. Yet sunspots, the visible manifestations of magnetic disturbances beneath the super-hot surface, have long tantalized scientists with possible relationships to conditions on Earth. The earliest recorded observations of sunspots are from China, dating to 28 BCE. In the early 1600s, Galileo and contemporaries began making extensive records of sunspots using the earliest telescopes. Starting in 1801 with the British astronomer Sir William Herschel (1738–1822), scientists proposed that solar variations could affect Earth's climate.

In the mid-nineteenth century an eleven-year, pulselike cycle of sunspot activity was identified. Then researchers discerned decades-long periods of sunspot quiescence and hyperactivity, called grand solar minima and maxima. The most famous such event is the Maunder Minimum—a remarkable period of solar calm from 1645 to 1720. Its scope was revealed in 1976 in a landmark paper in the journal *Science* by the astronomer John A. Eddy (1931–2009), who sifted an astounding array of evidence, from carbon isotopes in tree rings to records of past solar eclipses and sunspot patterns. He named the phenomenon for Edward Maunder (1851–1928) and Annie Maunder (1868–1947), the husband-and-wife team of astronomers whose meticulous work on ancient sunspot records, published in 1894, first pointed to the sunspot gap. (In a testament to the biases of the time, Annie Maunder's contribution to the work was not publicly recognized.)

The Maunder Minimum came during the centuries-long period known as the Little Ice Age and was initially thought to have been a substantial contributor to that cool spell. More recent research has shown that other factors tend to dominate.

Some scientists have proposed that changes in solar activity in the early twenty-first century may be the start of a new grand solar minimum (there have been only five in the last thousand years). With that in mind, in 2013 scientists at the National Center for Atmospheric Research examined whether a new sunspot drought could cool the planet enough to stop global warming.

The answer: Such a dip could slow warming, but not stop it in the long run.

**SEE ALSO:** Medieval Warmth to a Little Ice Age (1100), Space Weather Comes to Earth (1859)

---

**The sun has periods of high and low sunspot activity** that affect how much energy reaches the Earth. These two images, taken by NASA on October 28, 1998, and March 28, 2001, show the range, from quiet (bottom) to turbulent (top).

# FAHRENHEIT STANDARDIZES DEGREES

THERMOMETERS UTILIZING LIQUID IN A glass tube were developed in the 1630s. But the scientists and scholars using them had their own scales and often different reference points.

Without some common standard, there was no way to compare measurements at different places or times in a consistent way. Imagine a scientific observation—or even a cake recipe—without such a convention. Still, it took until the eighteenth century for standardization to emerge.

The inventor of the Fahrenheit scale was Daniel Gabriel Fahrenheit (1686–1736), who was the scion of a wealthy German merchant family in the Baltic port of Danzig (now Gdańsk, Poland). At age sixteen, after the death of both of his parents on the same day (by some reports, from unwittingly eating poisonous mushrooms), he was sent to Amsterdam to work for a shopkeeper. After spending four years at the shop, Fahrenheit became interested in making scientific instruments, including thermometers. He completed his first two alcohol thermometers in 1714 and set a temperature scale from 0 degrees (the temperature of a chilled brine solution) to 212 degrees (the temperature of boiling water).

In 1742, a Swedish astronomer, Anders Celsius (1701–44), was one of many scientists of this period to develop a 100-point temperature scale. What set his approach apart was his use of familiar benchmarks for both ends. He designated the freezing point of water as 100 degrees and the boiling point as 0 degrees. The two points were later switched to create the scale as it exists today. Celsius called his scale centigrade—Latin for one hundred steps. In 1948 most of the world adopted the Celsius scale as the standard unit of measurement for temperature.

The Kelvin scale is named after William Lord Kelvin (1824–1907), a Glasgow University engineer and physicist who, in a paper in 1848, proposed the need for a temperature scale starting at "infinite cold." The Kelvin scale is used mainly as a unit of temperature measurement in the physical sciences. Scientists use the Kelvin and Celsius scales simultaneously where absolute zero (0 K) is equivalent to –273 degrees Celsius.

**SEE ALSO:** The Invention of Temperature (1603), Settling a Hot Debate (2012)

**Early thermometer** designed by Daniel Gabriel Fahrenheit (1686–1736), made from brass, glass, and mercury, with a graduated scale from –4°F to 132°F.

# FOUR SEASONS ON FOUR STRINGS

THROUGHOUT HUMAN HISTORY, THE SOUNDS of the elements have influenced music and musical instruments, from the thunder of great drums to the windy notes of a Japanese flute to the drizzly rattle of rain sticks created from dried cactus by Chilean Indians. In Western classical music, the influence of meteorology was first fully articulated in *The Four Seasons*, a set of concertos composed around 1721 by the Italian violin virtuoso Antonio Vivaldi (1678–1741).

Each concerto conveyed the atmospherics of its time of year with chilly sprays or sleepy layers of notes. Although rarely included in performances now, a sonnet accompanied each one. Summer, for instance, including this line:

Soft breezes stir the air, but, threatening, the North Wind sweeps them suddenly aside.

It was Ludwig van Beethoven (1770–1827), however, who in 1802 pioneered a directly imitative approach to capturing the feeling of weather. In the fourth movement of his *Pastoral Symphony*, a thunderstorm approaches and builds explosively and then fades in what must have been an extraordinary experience for audiences at the time. Other composers quickly tried their hand. A 2011 paper in the British journal *Weather*, by Karen L. Aplin, a physicist at Oxford, and Paul D. Williams, an atmospheric scientist at the University of Reading, quantified the frequency of musical allusions to weather in classical music through the decades. The authors, both also classical musicians, found that storms were by far the most common phenomenon captured by composers.

Late in the nineteenth century, specialized instruments were developed to augment conventional orchestral instruments, including a metallic "thunder sheet" and wind machine, which generates a windlike *whoosh* with a revolving silk-covered drum.

With concerns rising about climate change from the buildup of greenhouse gases, in 2013, Daniel Crawford, a young cellist and geography student at the University of Minnesota, tried a novel sonification of climate data. He composed a piece for solo cello in which each note represents a year on NASA's record of global average temperatures from 1880 to 2012. He and others have composed more such music since.

SEE ALSO: China Shifts from Mythology to Meteorology (300 BCE), An Eruption, Famine, and Monsters (1816)

**Antonio Vivaldi** (1678–1741), pictured here with his violin, was one of the first Western composers to draw direct inspiration from meteorology.

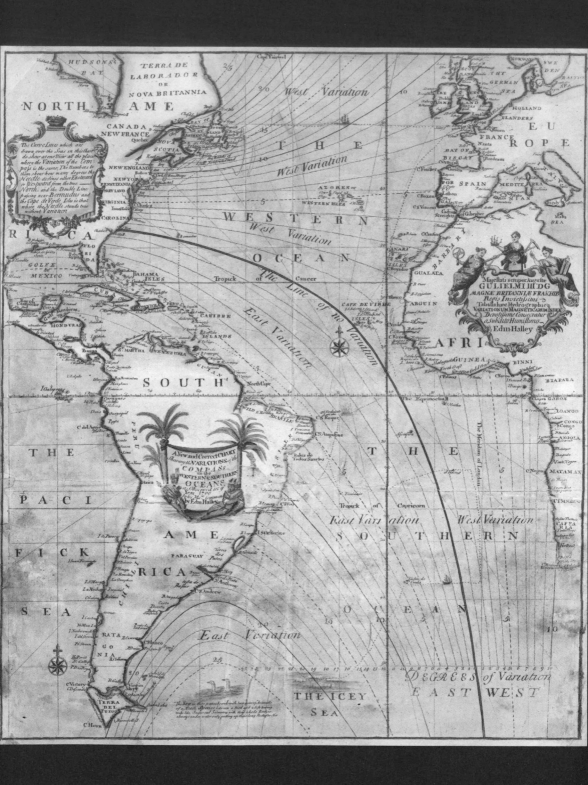

# MAPPING THE WINDS

EDMOND HALLEY (1656–1742), A SCIENTIST best known for his astronomical and mathematical work, took a stab at deciphering why the world has predictable winds in different geographic regions. In 1676, he sailed to Saint Helena Island in the remote South Atlantic to study the stars of the Southern Hemisphere, passing from northern temperate latitudes across the equator and southward. In 1686, Halley published an innovative global wind chart and a paper pointing to the heat of the tropics as the driver of the general motions of winds near and away from the equator. He proposed that the westward movement of air masses was related to the daily journey of the sun from east to west. After a colleague, the mathematician John Wallis (1616–1703), raised doubts about the idea that the sun could cause such wide patterns, Halley wrote that he began to doubt his own hypothesis.

In 1735, George Hadley (1685–1768), a lawyer and amateur student of weather, deciphered an important part of this meteorological puzzle. In a landmark paper, "Concerning the Cause of the General Trade Winds," he proposed, as did Halley's conception, that heated air rose and flowed poleward, then sank as it cooled, creating a looping cycle. But he also proposed the reason winds flowed at angles: The rotational velocity of the Earth is highest at the equator, so approaching air masses in essence fall behind the speed of the surface beneath. He wrote:

> From which it follows, that the Air, as it moves from the Tropics towards the Equator, having a less Velocity than the Parts of the Earth it arrived at, will have a relative Motion contrary to that of the diurnal Motion of the Earth in those Parts, which being combined with the Motion towards the Equator, a N.E. wind be produced on this Side of the Equator, and S.E. on the other . . .

Another century was required before other critical details in this complex system were clarified. But the basic looping atmospheric features became known as Hadley cells.

**SEE ALSO:** Beaufort Classifies the Winds (1806), Putting Wind to Work (1887)

**Drawing in part on observations** during voyages to and from Saint Helena Island in the South Atlantic Ocean in 1676, Edmond Halley produced a map of tropical trade winds a decade later that is recognized as a milestone in the history of cartography.

# BENJAMIN FRANKLIN'S LIGHTNING ROD

1752

BENJAMIN FRANKLIN (1706–90) IS BEST known as one of the Founding Fathers of the United States. He was also an author, printer, inventor, scientist, postmaster, diplomat, civic activist—and had a special fascination with early scientific studies of electricity. In 1747, Franklin began his own experiments, accidentally delivering a powerful shock to himself—"a universal blow throughout my whole body from head to foot," as he described the event in a letter.

Also a student of meteorology, he became convinced that lightning was similar to static electricity and started exploring ways to protect structures from this potent meteorological threat. In 1749, Franklin began developing the theory that a sharp-tipped iron rod, connected to the Earth, could shield a building from harm.

In June of 1752, he became impatient waiting for the completion of a church steeple in Philadelphia that he hoped to use as a test site for his lightning-rod concept. In the meantime, he undertook his legendary experiment of flying a kite in a thunderstorm with an iron key tied to the string. Franklin was lucky he survived this test, as some others who tried to repeat it later were electrocuted. As word of his work reached Europe, several experiments were conducted that confirmed his ideas.

Both the kite experiment and lightning-rod design demonstrated the scientific principle that electricity tries to find the path of least resistance to the ground. Drawing on these insights, Franklin published an article in the 1753 issue of his yearly *Poor Richard's Almanack* describing a method for protecting houses from lightning strikes. His system was made up of three key elements: a metal rod mounted at the peak of the roof, horizontal roof conductors, and vertical conductors to carry the charge to a ground connection.

Franklin erected a lightning rod on his own home, adding a novel detail—bells that would ring when the grounding wire became charged, notifying him when the air above the house was electrified. Franklin's lightning rods were eventually installed on a number of prominent structures, including the Pennsylvania State House, which later became Independence Hall.

SEE ALSO: Franklin Chases a Whirlwind (1755), Proof of Electrical "Sprites" (1989), Extreme Lightning (2016)

*Benjamin Franklin Drawing Electricity from the Sky* (c. 1816) by American-British painter Benjamin West (1738–1820) shows Franklin's famous kite experiment.

Plate V.        Vol. II. page

| 52 | 61 | 4  | 13 | 20 | 29 | 36 | 45 |
|----|----|----|----|----|----|----|----|
| 14 | 3  | 62 | 51 | 46 | 35 | 30 | 19 |
| 53 | 60 | 5  | 12 | 21 | 28 | 37 | 44 |
| 11 | 6  | 59 | 54 | 43 | 38 | 27 | 22 |
| 55 | 58 | 7  | 10 | 23 | 26 | 39 | 42 |
| 9  | 8  | 57 | 56 | 41 | 40 | 25 | 24 |
| 50 | 63 | 2  | 15 | 18 | 31 | 34 | 47 |
| 16 | 1  | 64 | 49 | 48 | 33 | 32 | 17 |

Fig.III. Page 326.

Fig.I. Page 26.

Fig.II. Page 26.

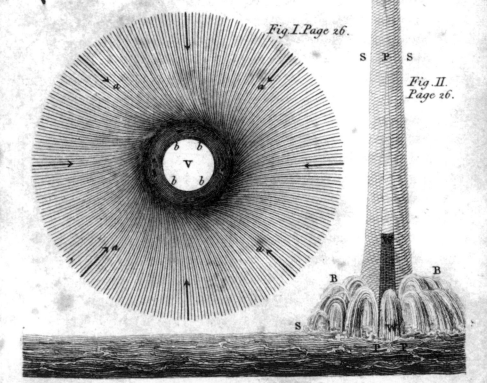

Published as the Act directs, April 2, 1806, by Longman, Hurst, Rees & Orme, Paternoster Row.

# FRANKLIN CHASES A WHIRLWIND

IN ADDITION TO HIS FOCUS ON LIGHTNING AND electricity, Benjamin Franklin had a long-standing fascination with tornadoes and other whirlwinds. This was evident in a series of letters and other writings on the subject, particularly in a detailed 1753 treatise on waterspouts, which included detailed diagrams laying out his theories on their anatomy and energetics.

But he clearly craved a closer look. In 1755, Franklin and his son William were staying at the Maryland estate of Colonel Benjamin Tasker. While out riding in the countryside, they came across a newly forming dust devil. Franklin recounted what happened next in a letter to Peter Collinson, a frequent correspondent on the topic of electricity. Here's an excerpt:

It appeared in the form of a sugar-loaf, spinning on its point, moving up the hill towards us, and enlarging as it came forward. When it passed by us, its smaller part near the ground, appeared no bigger than a common barrel, but widening upwards, it seemed, at 40 or 50 feet [12.2 or 15.2 meters] high, to be 20 or 30 feet [6.1 or 9.1 meters] in diameter. The rest of the company stood looking after it, but my curiosity being stronger, I followed it, riding close by its side, and observed its licking up, in its progress, all the dust that was under its smaller part. As it is a common opinion that a shot, fired through a water-spout, will break it, I tried to break this little whirlwind, by striking my whip frequently through it, but without any effect.

The chase ended when the whirlwind spun across a tobacco field and faded away, leaving the sky full of swirling leaves. Franklin ended his dispatch with this quip: "Upon my asking Colonel Tasker if such whirlwinds were common in Maryland, he answered pleasantly, 'No, not at all common; but we got this on purpose to treat Mr. Franklin.' And a very high treat it was. . . ."

SEE ALSO: Benjamin Franklin's Lightning Rod (1752), Beaufort Classifies the Winds (1806)

**Representation of a waterspout** that accompanied Benjamin Franklin's paper, "Water-spouts and Whirlwinds," republished in *The Complete Works in Philosophy, Politics, and Morals, of the Late Dr. Benjamin Franklin*, 1806.

# FIRST WEATHER BALLOON FLIGHT

ON NOVEMBER 21, 1783, THE FIRST manned balloon flight was taken by Jean-François de Pilâtre Rozier and the marquis d'Aalandes. It was an extraordinary scene, witnessed by Ben Franklin. In his journal, Franklin described the moment of ascent: "We observed it lift off in the most majestic manner. When it reached around 250 feet [76.2 meters] in altitude, the intrepid voyagers lowered their hats to salute the spectators. We could not help feeling a certain mixture of awe and admiration."

Before the flight, a weather observation balloon was launched to check the winds aloft, marking a lesser, but significant, milestone. In subsequent decades, weather balloons of increasing sophistication were used to reveal the structure and composition of the atmosphere. They still provide vital data to meteorologists today.

One of the pioneers in this pursuit was the French meteorologist Léon Teisserenc de Bort (1855–1913). In 1896, he discovered that above an altitude of seven miles (eleven kilometers), the temperature of the atmosphere remained relatively constant. In 1900, he concluded that the atmosphere is divided into two layers. Teisserenc de Bort called the lower atmosphere the *troposphere*, or "sphere of change." The troposphere not only contains most of the air and oxygen in our atmosphere, but it is also the layer where all weather takes place. Stratosphere was the name ascribed to the seemingly stable higher layer.

Even with all of our satellites and other monitoring systems, weather balloons are still launched twice a day—at precisely the same time—from 800 locations all over the world. They carry instrument packs called radiosondes some 18 miles (29 kilometers) high to gather information for forecasts. Readings on temperature, humidity, and pressure are transmitted by radio to ground stations. The data help generate weather maps or improve simulations generating forecasts.

Other kinds of weather balloons are used for research projects such as studying climate change or air pollution. Scientists also release weather instruments from balloons into storms so they can know the wind speed and direction of a given storm at different heights.

SEE ALSO: Franklin Chases a Whirlwind (1755), Putting Wind to Work (1887)

**Early testing for *radiosonde*** measurements involved hydrogen-filled balloons. A theodolite, pictured here, was used to track the balloon to the limit of visibility.

# THE FARMER'S ALMANAC

WHILE THERE HAVE BEEN MANY almanacs throughout the years, the best known—particularly for its gutsy long-range weather forecasts—is *The Old Farmer's Almanac* (originally *The Farmer's Almanac*), founded by Robert B. Thomas (1766–1846) and published without pause since 1792.

Thomas observed solar cycles and other phenomena to come up with a secret weather prediction model, which today is kept in a locked black tin box in the publication's offices in Dublin, New Hampshire. The magazine has long claimed to be 80 percent accurate in its predictions. In 1981, two atmospheric scientists, John Walsh and David Allen, published a different result in the magazine *Weatherwise*, assessing the accuracy of a substantial sample of *Farmer's Almanac* temperature and precipitation forecasts—and finding them hovering right around 50 percent.

The publication has long hinted at the value of not taking its findings too seriously. As Thomas wrote in 1829, the almanac "strives to be useful, but with a pleasant degree of humor."

The popularity of the almanac is said to have grown following a remarkably accurate prediction that appears to have been published by accident in 1816. According to Judson Hale, the longtime editor in chief of both today's *Farmer's Almanac* and *Yankee* magazine, in the edition that year, the forecast for January and February had been inadvertently transposed with July and August. The *almanac*'s founder desperately tried to recall all of the books, but news of that forecast had already gotten out. "He became the subject of much ridicule," Hale wrote, "until July brought rain, hail, and snow throughout New England!"

What happened was that Mount Tambora had erupted in Indonesia in 1815, blasting gases and dust high into the atmosphere, lowering global temperatures and indeed bringing summer snow to New England. A less-fortunate blunder came in the 1938 edition, in which the editor, Roger Scaife, replaced forecasts with temperature and precipitation averages. According to the *Almanac* website, "The public outcry was so great that he reinstated the forecasts in the next year's edition, but it was too late to save his reputation."

SEE ALSO: An Eruption, Famine, and Monsters (1816), Groundhog Day (1886)

---

**More than 200 years** after its founding, *The Old Farmer's Almanac* (originally *The Farmer's Almanac*) maintains an avid following despite, or perhaps because of, the "secret formula" behind its long-term weather predictions.

*Cirrus.*

*Cumul*

*Strah*

Lewis, sculp

# LUKE HOWARD NAMES THE CLOUDS

ON A DECEMBER EVENING IN 1802, LUKE Howard (1772–1864), a London pharmacist and amateur meteorologist, became the first person to propose a way to classify clouds. His ideas were presented to a small gathering of young science-minded intellectuals who called themselves the Askesian Society. Howard's lecture that evening was titled "On the Modification of Clouds" and opened as follows:

My talk this evening concerns itself with what may strike some as an uncharacteristically impractical subject: it is concerned with the modification of clouds. Since the increased attention which has been given to meteorology, the studies of various appearances of water suspended in the atmosphere has become an interesting and even necessary branch of that pursuit. If clouds were the mere result of condensation of vapour in the masses of the atmosphere which they occupy, if their variations were produced by the movements of the atmosphere alone, then indeed might the study be deemed a useless pursuit of shadows....

The talk, in essence, heralded the beginning of a formal science of meteorology. His presentation was published the following year as an essay, illustrated with many of his own sketches of cloud types, and appeared in subsequent publications over a span of almost twenty years. It is remarkable that Howard's classification, with minor changes, remains in use today, with meteorologists and the public at large still describing clouds with terms like *cirrus, cumulus,* and *stratus.*

Howard's classification was a revelation, bringing a sense of order and understanding to a subject that had lacked coordinated thought—let alone any documented theories as to how pressure, temperature, rainfall, and clouds might be related. Perhaps even more impressive was his intuition that clouds must be considered as "subjects of grave theory and practical research . . . governed by . . . fixed laws." Howard's ideas about the physics of clouds were generally sound, despite the poor understanding of the physics of air and water vapor in his time.

—*G. S.*

SEE ALSO: Beaufort Classifies the Winds (1806), First International Cloud Atlas (1896)

**One of Luke Howard's** drawings accompanying his pioneering descriptions of cloud types, which were published in 1803.

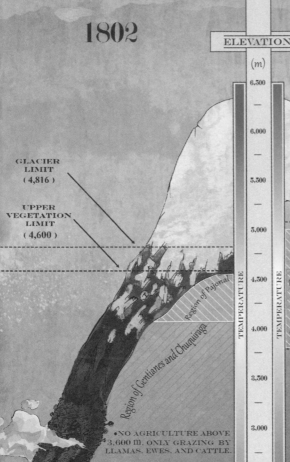

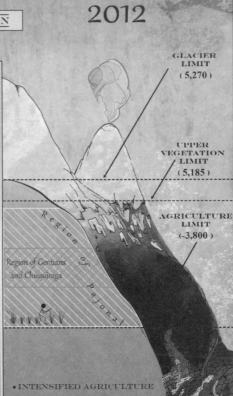

**1802**

**2012**

ELEVATION

(m)

6,500

6,000

5,500

5,000

4,500

4,000

3,500

3,000

2,500

TEMPERATURE

TEMPERATURE

REL. IMPACT

Δ LAND USE

Δ CLIMATE

GLACIER LIMIT ( 4,816 )

UPPER VEGETATION LIMIT ( 4,600 )

GLACIER LIMIT ( 5,270 )

UPPER VEGETATION LIMIT ( 5,185 )

AGRICULTURE LIMIT (~3,800 )

Region of Pajonal

Region of Gentianes and Chuquiraga

Region of Gentians and Chuquiraga

Region of Pajonal

• NO AGRICULTURE ABOVE 3,600 m, ONLY GRAZING BY LLAMAS, EWES, AND CATTLE.

• CULTIVATION OF POTATO ABOVE 3,000 m

• SCATTERED AGRICULTURE.

• INTENSIFIED AGRICULTURE AND LANSCAPE FRAGMENTATION UP TO 3,800 m

# HUMBOLDT MAPS A CONNECTED PLANET

AT THE DAWN OF THE NINETEENTH CENTURY, there were rumors that Napoleon Bonaparte envied the fame of only one man: Alexander von Humboldt (1769–1859). The roving Prussian naturalist and geographer had an extraordinary eye for local environmental details, global connections, and even potential disruption from the spreading activities of humans.

A year into an expedition across the Americas (1799–1804), he wrote in his diary about "mankind's mischief" disturbing "nature's order."

In 1802, Humboldt climbed more than 19,000 feet (5,791 meters) up the flank of Mount Chimborazo, a volcano in what was then called Gran Colombia but is now Ecuador, not only setting a mountaineering record but also gaining a critical insight about the interconnectedness of the living and physical worlds. Building on sketches there, he later drew a remarkable annotated side view of the mountain's climatic and ecological zones that some modern illustrators consider the first infographic.

More recently, in 2012, an international team of scientists replicated his ascent and analysis, producing—in the same style—a side-by-side comparison of his illustration's ecological zones and a modern variant showing the changes. Their study and the artwork were published in 2015 in the *Proceedings of the National Academy of Science*s, showing dramatic ecological shifts driven by the warming climate and encroaching agriculture.

Later in life, Humboldt, conceived of isotherms—the lines of common temperature still featured on climate charts. In helping establish a network of stations measuring Earth's magnetism, he became an evangelist for international scientific collaboration.

His travels and books inspired scientists and conservationists from Charles Darwin to John Muir. The importance of Humboldt to science and society can be conveyed best in the words of others, such as Darwin, who said his own voyage of discovery on the HMS *Beagle* was inspired by Humboldt's travels. Darwin wrote that Humboldt was "the greatest travelling scientist who ever lived," and added, "I worship him." Thomas Jefferson called him "one of the greatest ornaments of the age."

**SEE ALSO:** An Eruption, Famine, and Monsters (1816), First International Cloud Atlas (1896)

**In 2012, an international team** of scientists replicated Humboldt's survey of Mount Chimborazo, and in 2015, published a modern variant of his pioneering infographic alongside the original, showing climatic and ecological changes.

# BEAUFORT CLASSIFIES THE WINDS

1806

WHAT IS A GENTLE BREEZE, OR A GALE? There was no standard way to define wind strength until the early nineteenth century, when an Irish officer in the Royal Navy, Francis Beaufort (1774–1857), refined a straightforward way to describe wind speed as a function of the behavior of environmental features, from waves to tree branches, affected by moving air. The result was what is now called the Beaufort scale of wind velocity.

According to Britain's meteorological office, he devised the scale while serving aboard the HMS *Woolwich*. The first written record was his private log from January 13, 1806, stating he would "hereafter estimate the force of the wind" using this set of cues from nature. In 1807, he modified the scale somewhat, with the result being thirteen wind classes, numbered from 0–12. For example, a wind with a Beaufort number of 0 is described as calm, with speeds below 1 knot, or nautical mile per hour, resulting in a "sea like a mirror." As the Beaufort numbers increase, the wind speed rises along with descriptions of conditions. You wouldn't want to be at sea for number 12, hurricane force: "Huge waves. Sea is completely white with foam and spray. Air is filled with driving spray, greatly reducing visibility." The Beaufort scale can go up to 17, but numbers 13 through 17 only apply to tropical typhoons.

In 1916, with steam power rapidly supplanting sail, some descriptions of the behavior of canvas in the wind were dropped and others, adding more detail on sea conditions, were added. The scale was also adapted for use on land with the same thirteen classes but with descriptions of terrestrial responses to wind variations. At a number 1, "smoke rises vertically," and at 12, there is "considerable and widespread damage to vegetation, a few windows broken, structural damage to mobile homes and poorly constructed sheds and barns. Debris may be hurled about."

Meteorologists rarely use the Beaufort scale today, but it is still a good wind scale if no instruments are available.

SEE ALSO: Mapping the Winds (1735), Fastest Wind Gust (1934), Dangerous Downbursts Revealed (1975)

The **Beaufort scale of wind force** has thirteen levels defined using visual cues at sea or on land. The author, Andrew Revkin, encountered Force 7 conditions at the southern end of the Red Sea in 1980 while serving as first mate on a circumnavigating sailboat, the *Wanderlust*. As the scale notes, when winds reach a Beaufort number of 7, the "sea heaps up and white foam from breaking waves begins to be blown in streaks along the direction of the wind."

# LONDON'S LAST FROST FAIR

THERE ARE ACCOUNTS OF THICK ICE COVER-
ing the River Thames in London as long
ago as 250 CE. According to various
historical records, there were more than two
dozen winters between 1309 and 1814 when
ice on the Thames grew sufficiently thick to
support commerce of sorts.

In the coldest years, the result was a
sequence of "frost fairs"; that is, when the ice
became a playground, highway, and plaza—a
"carnival on the water," by one account from
that time. The ice could form more readily
because the river was wider than it is now, and
thus slower, and the nineteen arched passages
for river water beneath the Old London Bridge
tended to slow the flow further as they snagged
flotsam and ice floes.

The seven best-documented fairs took place
in cold winters between 1607 and 1814, in the
heart of a climatic period later called the Little
Ice Age. During the winter fair of 1607–08,
barbers, shoemakers, and other merchants
erected shops on the ice. Printing presses
cranked out souvenir cards. Oxen were roasted
and alcohol was heavily consumed.

During the brutally cold winter of 1683–84,
the Thames was frozen solid for two months.
The festival that year was dubbed the Blanket
Fair. A popular chronicler of the times, John
Evelyn, described the scene in his diary:

Coaches plied from Westminster to the
Temple, and from several other staires to
and fro, as in the streetes, sliding with skee-
tes, a bull-baiting, horse and coach races,
puppet plays and interludes, cookes, tipling
and other lewd places, so that it seemed a
bacchanalian triumph or carnival on the
water, whilst it was a severe judgement
on the land, the trees not onely splitting
as if lightning-struck, but men and cattle
perishing in divers places, and the very seas
so lock'd up with ice, that no vessels could
stir out or come in.

The climate grew milder and the Old London
Bridge was replaced in 1831 with a bridge that
allowed the water to flow more freely, making
the river less likely to freeze.

SEE ALSO: Medieval Warmth to a Little Ice Age (1100)

*Frost Fair on the Thames, with Old London Bridge in the distance* (c. 1685, artist unknown) shows Londoners gathered on the frozen Thames for a festival.

# AN ERUPTION, FAMINE, AND MONSTERS

IN THE SUMMER OF 1816, THE POET LORD BYRON (1788–1824) invited some literary friends to his home on the shore of Lake Geneva in Switzerland. But unusually gloomy cold rains settled in. To pass the time, Byron suggested they write horror stories. This is when Mary Shelley (1797–1851), a guest of Byron's, had a "waking dream" that would end up resulting in the timeless story of Frankenstein's monster. Lord Byron was inspired to write the poem "Darkness," which begins:

> I had a dream, which was not all a dream.
> The bright sun was extinguish'd,
> and the stars
> Did wander darkling in the eternal space,
> Rayless, and pathless, and the icy earth

These literary figures had no way of knowing that the gloomy weather was part of a globe-spanning climatic disruption triggered by a geological cataclysm that had occurred a year earlier and half a world away.

On April 10, 1815, Tambora, a long-simmering volcano in the Dutch East Indies (now Indonesia), blew its top. The blast was heard 1,500 miles (2,414 kilometers) away. Falling ash, tsunamis, and other local impacts killed thousands.

But the volcano also launched tens of millions of tons of sulfurous, sun-blocking particles into the atmosphere, creating a fast-spreading veil. The cool, wet weather experienced by Lord Byron and his guests had truly horrific consequences, also causing the failure of wheat, potato, and oat harvests in Ireland and England, resulting in the worst famine in Europe during the nineteenth century. In what became known as the Year Without a Summer or "Eighteen Hundred and Froze to Death," May frosts ruined crops in Massachusetts, Vermont, New Hampshire, and Upstate New York, and snow blanketed New England in June.

Scientists have estimated that up to 90,000 deaths worldwide could be attributed to the weather impacts from the sulfur cloud. The monsoon seasons in China and India were disrupted, resulting in flooding of the Yangtze River valley in China. In India, the late summer monsoon rains spread diseases like cholera.

There was one other impact in the arts. The conditions created by the volcanic aerosols shaped the paintings of the times, including a John Constable (1776–1837) landscape of a stormy coastline.

SEE ALSO: Dinosaurs' Demise, Mammals Rise (66 Million BCE), Four Seasons on Four Strings (1721), Nuclear Winter (1983)

*Weymouth Bay*, by English painter John Constable (1776–1837), depicts the dark, ashen skies on England's south coast after the eruption of Mount Tambora. Painted between 1819 and 1830, it was inspired by an earlier piece Constable painted on his honeymoon in 1816, the Year Without a Summer.

# WATERMELON SNOW

IN 1818, BRITISH NAVAL CAPTAIN SIR JOHN Ross (1777–1856) led an unsuccessful expedition aiming to find the fabled Northwest Passage over North America. Confused by complicated geography, he turned back. Sailing along Greenland's west coast, he made a far quirkier discovery. Distinctively pink snow could be seen on the white icy bluffs. Ross stopped and took samples of the snow, which he brought back to England—as water, of course. On December 4, 1818, the *Times* of London covered the discovery with some skepticism:

> Captain Sir John Ross has brought from Baffin's Bay a quantity of red snow, or rather snow-water, which has been submitted to chymical analysis in this country, in order to the discovery of the nature of its colouring matter. Our credulity is put to an extreme test upon this occasion, but we cannot learn that there is any reason to doubt the fact as stated.

Pink snow had been described by scholars as far back as Aristotle. But now scientific analysis was possible. Ross proposed the pink color was caused by debris from iron-rich meteorites, but a Scottish botanist, Robert Brown (1773–1858), suggested that species of algae were the culprit. Brown turned out to be correct.

What is now commonly called watermelon snow has since been found in snowy regions from Antarctica to Utah. The red snow condition appears in late spring and summer, as dormant algae spring to life when thin layers of meltwater form and are bathed in sunlight. The algae are green but produce red pigments as a kind of natural sunblock against harmful wavelengths of light.

The addition of color to the snow speeds melting, as sunlight that would reflect from a white surface is instead absorbed. A study published in 2016 by German and British scientists found that algae-darkened surfaces on glaciers in places like Greenland increase melting sufficiently that the effect needs to be considered in models used to predict the impacts of climate change.

**SEE ALSO:** Cold Dooms an Arctic Explorer (1845)

**Red snow**, also known as watermelon snow, caused by algae—in this case the flagellate *Chlamydomonas nivalis*—on the Tre Cime di Lavaredo ("three peaks of Lavaredo"), in the Dolomites Mountain Range in northern Italy.

*P.V. delin.*                    *J.Kent Fecit.*

## A MEETING of UMBRELLAS.

Pub.ᵈ Jan 25 1782 by W.Humphrey. 227 Strand.

# AN UMBRELLA FOR EVERYONE

FOR THOUSANDS OF YEARS, PEOPLE HAVE sought and refined ways to protect themselves from rain and sun while on the move. The oldest record of such a device is in ancient Egyptian hieroglyphs and carvings, showing royalty and gods with parasols over their heads. The first use of a waterproof counterpart, the umbrella, appears to have been in ancient China some 3,000 years ago. As in Egypt, such protection from the elements long remained a sign of royalty or nobility. Some Chinese emperors were shielded by elaborate four-tiered umbrellas. Artwork depicts rulers of Siam and Burma covered by umbrellas with up to twenty-four layers.

As the centuries passed, a variety of innovations made Asian-style parasols, and eventually the modern folding cloth-on-frame umbrella, more accessible to the masses.

In 1830, signaling the emergence of a bigger market, the first all-umbrella shop, James Smith & Sons, opened in London's West End. By the 1850s, umbrellas were finally emerging as a ubiquitous possession of British urbanites, largely thanks to the invention by Samuel Fox (1815–87) of lightweight steel ribbing that could substitute for the strips of whale baleen that had supported the silk or cotton fabric of earlier designs.

*Umbrellas and Their History*, a remarkable book by William Sangster, published in Great Britain in 1855, amusingly chronicled the umbrella's transition from oddity to essential:

> Only a few years back those who carried Umbrellas were held to be legitimate butts. They were old fogies, careful of their health, and so on; but now-a-days we are wiser. Everybody has his Umbrella. It is both cheaper and better made than of old; who, then, [is] so poor he cannot afford one? To see a man going out in the rain umbrella-less excites as much mirth as ever did the sight of those who first—wiser than their generation—availed themselves of this now universal shelter.

SEE ALSO: "Manufactured Weather" (1902), The Windshield Wiper (1903)

**A Meeting of Umbrellas** (1782) by British cartoonist James Gillray (1756–1815). At that time, the sight of a gentleman carrying an umbrella suggested excessive dandyism or effeminacy.

# ICE AGES REVEALED

PONDERING ALPINE LANDSCAPES, EARLY scholars in the West were baffled by great scratchlike marks on bedrock and giant boulders that looked like they had been randomly tossed by giants. Through the eighteenth century, the dominant theory had been that these landscape features were evidence of ancient floods, possibly the biblical deluge. Then a new idea slowly emerged. In 1815, a Swiss hunter and mountaineer, Jean-Pierre Perraudin, concluded that the giant granite rocks scattered in valleys in the Alps had to have been carried down by advancing glaciers. He explained his observation to an engineer, Ignaz Venetz (1788–1859), who began mapping such features and gave a lecture on the concept in 1829. But he was met with deep resistance.

Jean de Charpentier (1786–1855), a German-Swiss geologist who had met Perraudin in 1815, later recalled how he found the hypothesis so extravagant at first that he deemed it "not worth examining or even considering." But he came around, partially through pressure from Venetz.

It took one more step to begin moving the concept from quirky proposition to new paradigm. Charpentier, who had compiled maps and data pointing to such glacial dynamics, ended up convincing Louis Agassiz, a prominent Swiss zoologist and geologist, to weigh the evidence. Agassiz, like others, resisted at first, but then took hold of the idea and gave a memorable lecture in 1837 and wrote a book in 1840, *Études sur les Glaciers* (Glacier Studies), that introduced the phrase *ice age* to science. Later observations of the great impact of ice sheets in North America reinforced and expanded his view.

Doubts persisted into the 1870s. But the scientific picture crystallized. Toward the end of the century, geologists found evidence showing that ice sheets had advanced and retreated at least four times, with each cycle spanning tens of thousands of years. The next challenge was figuring out what was driving the changes.

SEE ALSO: Orbits and Ice Ages (1912), Climate Clues in Ice and Mud (1993)

**Illustration from Louis Agassiz's 1840 atlas**, *Études sur les Glaciers*, showing a glacier in Zermatt, Switzerland.

# PEAT BOG HISTORY

THERE WASN'T MUCH THAT THE DANISH naturalist Japetus Steenstrup (1813–97) didn't find of interest. He lectured on mineralogy, studied the sexuality of worms, showed that the fabled kraken was just an oversize squid, critiqued Stone Age carvings, loaned much of his vast barnacle collection to Charles Darwin, taught microscopy to the man who discovered the Gram's stain technique for bacteria, and taught botany to the future founder of plant ecology. And, in 1836, he dug up a couple of bogs.

Steenstrup carefully recorded the different layers of peat (old decomposed vegetation), identified the plant fossils in those layers, and showed that the plant species growing in and around the bogs had changed over the years. He reasoned that climate change drove those transitions, and developed the world's first sediment-based climate chronology for conditions since the last ice age.

His discoveries came on the shoulders of work by George Cuvier (1769–1832) and James Hutton (1726–97), who demonstrated great changes in the deep history of the Earth, and about the same time as Louis Agassiz (1807–73) was proposing glacial theory and Charles Lyell (1797–1875) was bringing a systematic approach to geology.

Steenstrup's 1841 paper on his bog studies demonstrated that climate and vegetation had shifted substantially over thousands of years, and represented the dual dawning of the modern sciences of paleoclimatology and paleoecology—studies of evidence of past climate and ecological change. His work inspired more refined peat-bog studies in Scandinavia in the following decades, leading directly to: the Blytt-Sernander climate sequence, a classification of northern European climatic phases; the discovery of an abrupt short-lived climate change called the Younger Dryas; and the 1916 development of methods for analyzing old pollen as a clue to past climates and ecological patterns. Such pollen analysis reveals rates and magnitudes of natural changes to compare with current and projected changes. It all began with a curious naturalist in a boggy landscape.

—S. J.

SEE ALSO: Shen Kuo Writes of Climate Change (1088 CE), Ice Ages Revealed (1840)

Layers of peat can be stripped away and used for a number of purposes, from fuel to whisky production. Peat bog studies have also produced insights into thousands of years of regional changes in climate and ecology.

# COLD DOOMS AN ARCTIC EXPLORER

In 1845, Sir John Franklin (1786–1847), a British explorer who gained fame after leading two expeditions in northernmost North America, made the mistake of heading toward the Arctic a third time. He had earned the nickname "the man who ate his boots" after surviving a particularly grueling trek that killed most of his companions. So it was with much fanfare that Franklin left England in the spring of that year on a mission to find the Northwest Passage, the water route that was thought to connect the Atlantic and Pacific Oceans.

He departed with a crew of 134 aboard two sailing ships, the HMS *Erebus* and the HMS *Terror*. The reinforced vessels had already successfully probed icy waters around Antarctica. They were fitted with steam engines and propellers that could be raised or lowered into the sea to help push through ice. In late June, some whalers in Baffin Bay, west of Greenland, spotted the ships tethered to an iceberg. That was the last time they were seen by anyone other than the Inuit natives of the region.

Through more than a decade, a string of expeditions set out from Britain and the United States, trying to retrace Franklin's route in hope of finding the lost crews and their leader. In the end, all that was found were traces of abandoned camps and Inuit stories and—in 1851—the graves of three of the crew. But those voyages, some forty in all between 1848 and 1859, resulted in a great burst of Arctic exploration and science.

Many factors appear to have contributed to the loss of the ships and their crews, with theories of lead poisoning from poorly sealed tins of food largely debunked by 2016. Recent research concluded that bad climatic luck played a significant role. In the decade of Franklin's last journey, that part of the Arctic was at its coldest point in many centuries.

It took until 2014 and 2016 for the sunken remains of the *Erebus* and the *Terror* to finally be discovered, in waters so routinely free of sea ice in summer that luxury cruise ships now unremarkably pass through that same passage that the nineteenth-century chill denied to Franklin.

SEE ALSO: Medieval Warmth to a Little Ice Age (1100), Russia's "General Winter" (1941), The Polar Vortex (2014)

**This 2006 photograph, taken on Beechey Island** in the Canadian Arctic, shows the tombstones of some of the crew of the ill-fated expedition led by British Arctic explorer Sir John Franklin.

Stamped Edition, 6d.

# THE ILLUSTRATED
# LONDON NEWS

ONE PENNY

REGISTERED AT THE GENERAL POST-OFFICE FOR TRANSMISSION ABROAD.

No 1594.—VOL. LVI.　　　SATURDAY, MAY 14, 1870.　　　WITH A SUPPLEMENT, FIVEPENCE } STAMPED, 6D.

PROFESSOR TYNDALL LECTURING AT THE ROYAL INSTITUTION.
SEE PAGE 510.

# SCIENTISTS DISCOVER GREENHOUSE GASES

IN 1824, THE FRENCH MATHEMATICIAN AND physicist Joseph Fourier (1768–1830) became the first person to propose how the atmosphere could mediate Earth's climate by allowing the passage of solar energy into the atmosphere as visible sunlight, but impeding the escape of invisible radiant heat energy back to space. It took more than thirty years before scientists developed the capacity to confirm this and discern which atmospheric gases were at work.

At a scientific conference in Albany, New York, in August 1856, Eunice Foote (1819–88), an American amateur scientist and advocate for women's rights, reported the results of some simple experiments on the contribution of carbonic acid (carbon dioxide) and moist air (water vapor) to the heating power of sunlight. She had to do so through a proxy because women could not speak at the conference. In the related paper, published in November, she observed: "The highest effect of the sun's rays I have found to be in carbonic acid gas," and added, "an atmosphere of that gas would give to our earth a high temperature."

In May 1859, unaware of Foote's work, a prominent Irish scientist, John Tyndall (1820–93), began using an instrument of his own design, a ratio spectrophotometer, to measure how various gases absorbed and radiated energy. Tyndall reported finding enormous differences, with the most abundant constituents of air—nitrogen and oxygen—essentially transparent to heat, while rarer gases, most notably water vapor and carbon dioxide, had a powerful effect.

In 1861, he articulated how variations in the abundance of these gases "may have produced all the mutations of climate which the researches of geologists reveal." In the course of his experiments, Tyndall also realized that cities could have a local heating impact on temperature. He coined the term *heat island*—a concept now well established by subsequent research.

By century's end, other scientists would begin to gauge the climate impact of a big buildup of carbon dioxide that was occurring through the burning of vast amounts of coal, with optimistic interpretations at first and then growing concerns.

SEE ALSO: Coal, $CO_2$, and the Climate (1896), The Rising Curve of $CO_2$ (1958), Climate Models Come of Age (1967)

**Irish-born physicist John Tyndall** lecturing on electromagnetism at the Royal Institution, London, in May 1870.

# SPACE WEATHER COMES TO EARTH

I N THE MID-NINETEENTH CENTURY, WITH industrializing nations relying increasingly on communication through telegraph lines, an explosive disturbance on the sun demonstrated that humanity would subsequently have to worry about extraterrestrial weather, not just about tempestuous conditions in the atmosphere on Earth.

The first sign something extraordinary was unfolding came late in the morning on September 1, 1859, in Britain. An astronomer, Richard C. Carrington (1826–75), was doing his regular documentation of sunspots by watching an image of the sun projected from his telescope onto a glass plate. Sunspots are (relatively) cool regions on the sun's surface reflecting distortions of the magnetic field generated within and around the turbulent sphere of super-hot ionized gases.

At 11:18 a.m., as he studied an unusually broad sunspot array, he was startled to see two pinpricks of bright light appear over the dark area. The phenomenon faded before he could find a witness. But luckily another astronomer, Richard Hodgson (1804–72), had independently recorded the event.

In the following hours, much of the world witnessed the consequences of what was the strongest solar flare ever recorded by science, then or since. The flare buffeted Earth's outer atmosphere with X-rays and ultraviolet radiation traveling near the speed of light. Lagging by half a day, a slower "coronal mass ejection"— billion-ton clouds of high-energy plasma— surged past the planet.

Chaos erupted in telegraph systems worldwide. Sparks jumping from wires shocked operators and set telegraph paper on fire. Luminous multicolored auroras, the shimmering curtains of electron-charged northern nights normally restricted to high latitudes, lit up the skies from the Caribbean to Hawaii.

It's just a matter of time before another such event occurs. NASA reported a near miss in July 2012. The cost of a direct blow to the world's electronics-dependent economy could top $2 trillion, according to studies by the National Academy of Sciences. In 2015, the Obama White House released the nation's first National Space Weather Action Plan, calling for a host of steps to boost preparedness and limit risks.

SEE ALSO: A Spotless Sun (1645), Watching Weather from Orbit (1960)

*Aurora Borealis*, painted in 1865 by American artist Frederic Edwin Church (1826–1900), was inspired by sketches of the northern lights made in 1860 by the Arctic explorer Isaac Israel Hayes. In 1859, a solar storm spread such atmospheric disturbances as far south as the Caribbean.

# FIRST WEATHER FORECASTS

ROBERT FITZROY (1805–65) IS REMEMBERED mainly as Charles Darwin's captain on the legendary circumnavigation by the HMS *Beagle* in the 1830s. But FitzRoy found fame, and sometimes criticism, in his own right for creating the first daily weather predictions, to which he gave the name *forecasts*.

FitzRoy was haunted by the unending loss of sailors' lives as vessels sank in storms along Britain's coasts. Between 1855 and 1860 alone, 7,402 ships were wrecked and more than 7,000 lives lost at sea. FitzRoy was convinced that by conveying better weather insights, he could reduce that toll. After the disastrous sinking of the *Royal Charter* clipper in 1859, with the loss of 450 lives, he was given the authority to start a storm warning service for mariners in February 1861.

The key to making forecasts was the telegraph. FitzRoy, who had also devised some new variants of the barometer, began building Britain's first weather office, fed by a network of weather observers whose reports would allow him to chart and predict the movements of storm systems. As the network expanded, the potential to get ahead of the weather did, as well. When he calculated that a certain port was at risk, FitzRoy could telegraph local officials. According to the British Broadcasting Corporation (BBC), FitzRoy characterized forecasting as a "race to warn the outpost before the gale reaches them."

The first public weather forecast for the common citizenry was published in the *Times* of London on August 6, 1861. It was rather basic:

North—Moderate westerly wind; fine.
West—Moderate south-westerly; fine.
South—Fresh westerly; fine.

The following year, a system was introduced in which cone-shaped signals were raised at major ports when a gale was predicted. FitzRoy capped his career by writing *The Weather Book: A Manual of Practical Meteorology,* which was published in 1863.

**SEE ALSO:** Beaufort Classifies the Winds (1806), Meteorology Gets Useful (1870), Tornado Warnings Advance (1950)

**Portrait of Robert FitzRoy**, English navigator and meteorologist, by English painter Samuel Lane (1780–1859).

K, STREET, FROM THE LEVEE.

INUNDATION OF THE STATE CAPITOL,

City of Sacramento, 1862.

Published by A ROSENFIELD, San Francisco.

# CALIFORNIA'S GREAT DELUGE

I N 1861, CALIFORNIA FARMERS AND RANCHERS were praying for rain after two exceptionally dry decades. In December, their prayers were answered with a vengeance, as a series of monstrous Pacific storms slammed into the west coast of North America, from Mexico to Canada. More than five feet (1.52 meters) of rain fell in Los Angeles in a year—more than four times the average—causing rivers to surge over their banks, spreading muddy water for miles across the arid landscape.

Early in 1862, the enormous pulse of water created a huge inland sea in California's Central Valley, covering a region 300 miles (482.8 kilometers) long and 20 miles (32 kilometers) wide. Water covered farmlands and towns, drowning people, horses, and cattle, and washing away houses, buildings, barns, fences, and bridges. The water reached depths up to 30 feet (9.1 meters), completely submerging telegraph poles that had just been installed between San Francisco and New York. This caused transportation and communications to completely break down over much of the state for a month. Botanist William Henry Brewer (1828–1910) wrote a series of letters to his brother on the East Coast describing the surreal scenes of tragedy that he witnessed during his travels in the region that winter and spring. In a description dated January 31, 1862, Brewer wrote:

> The entire valley was a lake extending from the mountains on one side to the coast range hills on the other. Steamers ran back over the ranches fourteen miles [23 kilometers] from the river, carrying stock, etc, to the hills. Nearly every house and farm over this immense region is gone. America has never before seen such desolation by flood as this has been, and seldom has the Old World seen the like.

The waters subsided, but the risk of such events persists. Research has since shown how certain conditions in and over the Pacific Ocean periodically fuel this kind of system, now called atmospheric rivers. They remain hard to predict, and it's not clear whether global warming will worsen them. But one thing is not hard to predict—there will be enormous financial losses the next time, with one study estimating the cost at more than $700 billion.

—L. I.

SEE ALSO: Midwestern Firestorms (1871), The Human Factor in Weather Disasters (2006)

**Lithograph of K Street** in the city of Sacramento, California, during the great flood of early 1862, which turned much of the state's Central Valley into a vast lake.

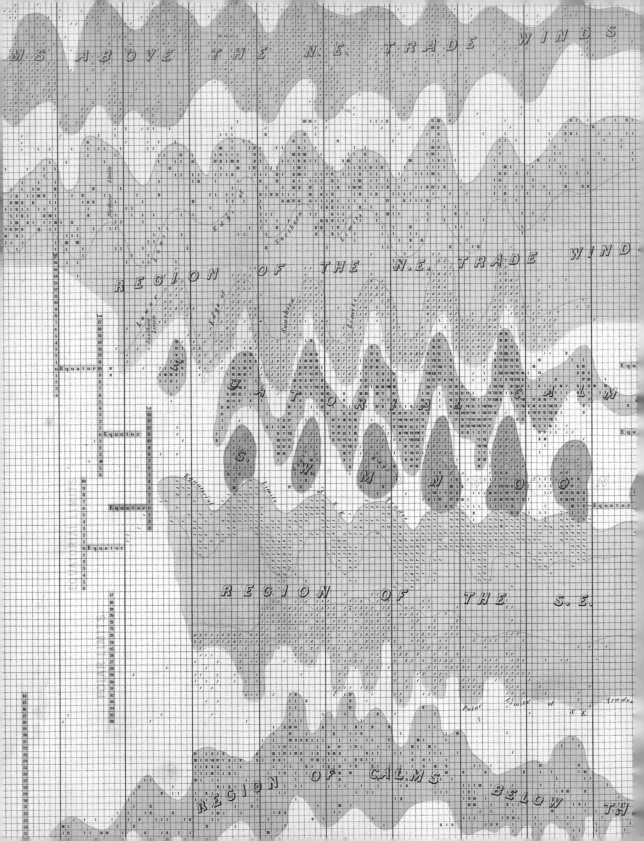

MS ABOVE THE N E TRADE WINDS

REGION OF THE N.E. TRADE WINDS

Northern Limits

Lower Southern Limit

Edge of

Northern Limits

Southern

Limit

EQUATORIAL CALMS

Equator

Equatorial Limits

Edge of S.E. Trade

Equator

S.W. MONSOON

Equator

REGION OF THE S.E.

Polar Limit of S.E. Trades

REGION OF CALMS BELOW THE

# METEOROLOGY GETS USEFUL

METEOROLOGY WENT THROUGH A momentous transition in the mid-nineteenth century, transforming from a mainly scholarly or informal pursuit to an organized and vital public service underpinning everything from agriculture to public safety, navigation to military preparedness.

The first focus of this emerging science was at sea, where weather, like nowhere else, was a matter of life or death and where winds could determine the outcome of both commerce and war.

Matthew Fontaine Maury (1806–73), an American naval officer with a passion for science, developed standardized methods for recording atmospheric and oceanic conditions. In 1853, he organized the first International Maritime Meteorology Conference in Brussels, Belgium. This resulted in standardizing weather-reporting practices in thirteen countries. Maury's maps of ocean wind and current patterns remain a marvel today. A string of terrible shipwrecks in Great Britain propelled forecasting efforts there.

The modern U.S. National Weather Service is built on a foundation created on February 2, 1870, when President Ulysses S. Grant signed a congressional resolution authorizing the secretary of war to establish an agency "to provide for taking meteorological observations at the military stations in the interior of the continent and at other points in the States and Territories . . . and for giving notice on the northern [Great] Lakes and on the seacoast by magnetic telegraph and marine signals, of the approach and force of storms."

That November, the first coordinated meteorological reports were taken by observers at twenty-four stations and transmitted by telegraph to Washington, D.C. A school of meteorology was added to an existing school of telegraphy and military signaling at an army post near Arlington National Cemetery in Virginia.

By 1873, thousands of rural post offices were receiving forecasts and posting "Farmers' Bulletins," augmented in 1881 by a system of signal flags with different patterns and colors for each weather condition. (A cold-wave flag had a white field with a black square at the center.) The weather service became a civilian agency in 1890.

SEE ALSO: First Weather Forecasts (1861), Tornado Warnings Advance (1950)

Trade wind chart of the Atlantic Ocean by Matthew Fontaine Maury, from 1851, compiled by Lt. E. J. Dehaven from materials in the U.S. Bureau of Ordnance and Hydrography.

# MIDWESTERN FIRESTORMS

THE WORST WILDFIRE IN NORTH AMERICAN history, in both extent and loss of lives, swept through northeastern Wisconsin and the upper peninsula of Michigan, from October 8–14, 1871, killing between 1,200 and 2,400 people and consuming woods and towns across 3.8 million acres (15,378 square kilometers). But that disaster, called the Great Peshtigo Fire, is largely unknown outside of a small circle of historians because of another conflagration that erupted at the same time—the Great Chicago Fire. This blaze killed some 300 Chicagoans as wind-whipped flames surged from the south through a city largely comprised of wooden structures. A colorful legend—that the fire started when Mrs. O'Leary's cow kicked over a lantern—has given that disaster story extra longevity. A third great fire in Michigan burned some 2.5 million acres (10,117 square kilometers).

A variety of theories emerged seeking some common cause, with the breakup of a meteorite considered, but mostly rejected. Meteorologists point to a potent summer drought, a week of high winds, and the prevalence of the deliberate use of fire as a land-clearing method in the fast-growing region as the most likely factors.

An eyewitness to the Wisconsin fire, the Reverend Peter Pernin of the Peshtigo Catholic Parish, certainly added credence to this scenario. He described how farmers and railroad workers routinely used "both axe and fire to advance their work," and how the days ahead of the great blaze saw many spreading fires.

In a chilling account of the unfolding tragedy, published a century later in 1971 by the State Historical Society of Wisconsin, Pernin described how terrified townsfolk gathered by the river and had to immerse themselves to survive:

> The banks of the river as far as the eye could reach were covered with people standing there, motionless as statues, some with eyes staring, upturned towards heaven, and tongues protruded. I pushed the persons standing on each side of me into the water. One of these sprang back again with a half smothered cry, murmuring: 'I am wet'; but immersion in water was better than immersion in fire.

SEE ALSO: The Dust Bowl (1935), Extreme Lightning (2016)

**Drawing by G. J. Tisdale**, 1871, of the Great Peshtigo Fire, which ultimately killed 1,200 to 2,400 people. Here, frantic residents seek refuge in the Peshtigo River.

# "SNOWFLAKE" BENTLEY

I S IT REALLY TRUE THAT EVERY SNOWFLAKE IS unique in its flowery crystalline shape? Recent science has confirmed this for fully formed flakes. But early inspiration for the idea that no two snowflakes were alike came from the obsession with all things small, and cold, of a boy raised on a Vermont farm in the second half of the nineteenth century. The boy, Wilson Alwyn Bentley (1865–1931), became known as "Snowflake" Bentley after developing a life-long passion for photographing snow crystals through a microscope. His thousands of images of snowflakes helped prompt research into the formation of frozen precipitation.

Bentley was homeschooled by his mother, who had been a teacher, until the age of fourteen. Thanks to his mother's microscope, he became fascinated with minutiae. As he recalled in an interview for *The American Magazine* in 1925, "When the other boys of my age were playing with pop guns and slingshots, I was absorbed in studying things under this microscope: drops of water, tiny fragments of stone, a feather from a bird's wing. But always, from the very beginning, it was the snowflakes that fascinated me most."

After reading about ways to take pictures through a microscope, Bentley and his mother convinced his father to buy him a suitable camera. In his later writings, he noted this was no small investment, with the camera costing $100 even then.

He then refined his technique. After catching a flake, he would use a feather to position it under the lens. Every step had to be done outdoors in the cold so the snow crystals wouldn't melt during the long exposure required to capture the image.

Shortly before his death, Bentley collaborated with William J. Humphreys (1862–1949), a physicist with the U.S. Weather Bureau, to produce *Snow Crystals*, a book containing 2,300 of Bentley's photographs. The book is still in print today. (The Weather Bureau became the National Weather Service in 1970.)

By the time of his death—from pneumonia—at the age of sixty-six, Bentley had photographed more than 5,000 snow crystals, each one indeed unique.

SEE ALSO: Luke Howard Names the Clouds (1802), Watermelon Snow (1818), The Great White Hurricane (1888)

**Snow crystal photographs** by Wilson Alwyn Bentley, taken in the 1890s through 1920s.

# COORDINATING ARCTIC SCIENCE

NEARLY A CENTURY BEFORE THE FIRST spacecraft offered humans a wide-angle view of their home planet, researchers from a dozen nations took part in the International Polar Year of 1882–83, attempting to create the first comprehensive analysis of meteorology around the Arctic. This revolutionary effort to understand the remarkable conditions at high latitudes probed everything from the brutal weather to the magnetic fields, from the frozen ocean to the shimmering aurora borealis (or, northern lights). The research stations had no way to stay in touch with one another. They collected data in isolation from the outside world month after month. But the instrument calibrations and recording methods at each station were coordinated, so that once the project was done, scientists could compile all the information and create the first snapshot of the remote region. A smaller effort was made in areas near Antarctica.

The Polar Year project grew out of the ideas of Karl Weyprecht (1838–81), an Austrian explorer and physicist. After completing an expedition in the Arctic in 1874, he traveled from one scientific organization to another, urging them to create a unified research mission. The international push into the Arctic until then, he wrote in an 1875 report, "The Principles of Arctic Exploration," had amounted to little more than a dangerous contest. As he wrote: "Immense sums were being spent and much hardship endured for the privilege of placing names in different languages on ice-covered promontories, but where the increase in human knowledge played a very secondary role."

Altogether, fourteen stations were established around the Arctic by a dozen nations: the Austro-Hungarian Empire, Denmark, Finland, France, Germany, the Netherlands, Norway, Russia, Sweden, the United Kingdom, Canada, and the United States. The effort was fraught with difficulties and danger. An American expedition sent twenty-five men into the Arctic to collect data—only six returned. Weyprecht died the year before the project got underway, but he is now recognized as having inspired a great moment in science—the first time some portion of Earth was probed by researchers from many countries and disciplines in the pursuit of shared knowledge.

SEE ALSO: Cold Dooms an Arctic Explorer (1845), Meteorology Gets Useful (1870)

**A Dutch Arctic expedition** sets up camp in 1883 as part of the first International Polar Year.

# FIRST PHOTOGRAPHS OF TORNADOES

From Benjamin Franklin's horseback pursuit of a whirlwind to today's live-streaming "storm chasers," weather watchers have been fixated with gaining a close-up view of tornadoes and documenting what they see. As the use of photography grew late in the nineteenth century, it was only a matter of time before the first images of tornadoes were captured and conveyed to a fascinated public.

It was long thought that the first tornado photograph was taken on August 28, 1884, as four powerful twisters formed near Howard, in the southeast corner of what was then the Dakota Territory. The outbreak claimed at least six lives, which is probably why the photograph taken that day by F. N. Robinson garnered the most attention. The image is terrifying, showing a central black funnel kicking up a large cloud of debris and flanked on each side by smaller, hornlike vortices. As was common at the time for unusual or newsworthy imagery, Robinson's photograph was reproduced on souvenir cards.

But weather historians have since concluded that a photograph taken in Garnett, Kansas, by A. A. Adams on April 26, 1884, was most likely the first—albeit of a less dramatic storm. The photo by Adams shows a ropelike tornado that was probably close to dissipating. (Given how long it would have taken to set up one of the cumbersome cameras of the day and record an image, that's no surprise.) Adams, like Robinson, sold souvenir cards, and also stereographs, which were side-by-side images viewed through a stereoscope that provided a three-dimensional feel.

The most concerted effort to clarify the sequence of these early tornado photographs was made by Purdue University meteorologist John T. Snow in a 1984 paper for the *Bulletin of the American Meteorological Society*. While Snow concluded that "a definitive statement concerning the identification of the first tornado photograph probably can never be made," he offered evidence from a variety of sources bolstering Adams as the pioneer.

**SEE ALSO:** Franklin Chases a Whirlwind (1755), Storm Chasing Gets Scientific (1973)

**Several nineteenth-century photographs of tornadoes** were said to be the first such images, but recent research gave that honor to this photograph, taken near Central City, Kansas, on April 26, 1884, by A. A. Adams.

# GROUNDHOG DAY

EVERY FEBRUARY 2, IN A CEREMONY THAT begins well before sunrise, tens of thousands of people from around the globe gather at Gobbler's Knob in Punxsutawney, Pennsylvania, to await the spring forecast from a very special groundhog (a rodent also known as a woodchuck). Called Punxsutawney Phil, this weather-forecasting creature is pulled from a mock tree-trunk home, huddled over mysteriously by top hat–wearing officials, and held aloft to cheers, in a ceremony with roots in ancient lore.

Groundhog Day marks the halfway point between the winter solstice and spring equinox. Early Christians, possibly drawing on an earlier Pagan tradition, had a winter weather forecasting custom associated with Candlemas Day. According to an old Scottish couplet, "If Candlemas is bright and clear, There'll be twa [two] winters in a year." In a German variant, if a hedgehog cast a shadow on Candlemas Day, there would be six more weeks of winter, or a "second winter."

German settlers in Pennsylvania began to use the groundhog, which they also hunted and ate with gusto in summer, for the ritual. The Punxsutawney groundhog forecast became official on February 2, 1886. On that day, Clymer H. Freas, city editor of the local paper, declared that Phil, the Punxsutawney groundhog, was America's only true weather-forecasting groundhog. Locals insist it's been the same groundhog ever since, sustained on a special elixir. (The species actually only live an average of six or seven years.)

How accurate are Phil's forecasts? The Punxsutawney Groundhog Club, which takes care of Phil, has kept records of the forecasts since the first Gobbler's Knob procession, in 1887. In 2017, the PennLive.com website ran an article summarizing its thorough analysis of predictions and winter weather, concluding, "In 117 years of available records, our calculations show Phil and his translators have been correct about 65 percent of the time."

Members of the club blame any mistakes on human error. In the PennLive article, the club's groundhog handler, Ron Ploucha, explained, "Unfortunately, there have been years where the [club] president has misinterpreted what Phil said." He added, "Phil's never wrong."

SEE ALSO: China Shifts from Mythology to Meteorology (300 BCE), *The Farmer's Almanac* (1792)

**Punxsutawney Phil** has been prophesying from his Pennsylvania home since the late 1800s.

# PUTTING WIND TO WORK

THE WIND HAS BEEN PUT TO WORK FOR thousands of years, first through the use of sails to propel ships, from the Nile to the Aegean to the Pacific. Persian farmers used wind power to pump water and grind grain between 2,500 and 3,000 years ago. Wind power spread in the Middle East and Europe. From the Medieval period onward, windmills became a particularly iconic presence in the Netherlands, where the devices had many uses, including draining marshes in the soggy lowlands.

But a critical leap came in the 1880s, when the first electrical generators powered by wind turbines were built on both sides of the Atlantic. In 1887, James Blyth (1839–1906), a Scottish professor and engineer, experimented with three different designs for wind-powered generators and installed a small one to light his vacation home.

Particularly remarkable was the 40-ton, 12,000-kilowatt wind turbine built by the engineer and inventor Charles F. Brush (1849–1929) on the grounds of his Cleveland mansion in the winter of 1887–88. Brush had already grown wealthy from inventing the electricity-generating dynamo and arc light systems that by 1881 were lighting up nights in cities from Boston to San Francisco and abroad. The firm he founded, Brush Electric Company, eventually merged with other companies to become General Electric.

Brush's wind-powered generator was enormous, with a 50-foot-diameter (15.2-meter) windmill wheel and 144 blades. The system powered the mansion continuously for twenty years.

In the twentieth century, the rapid expansion of coal-burning plants and then natural gas killed interest in wind. But the energy crisis of the 1970s triggered a revival that has accelerated along with concerns about pollution and climate change.

As of 2017, more than 240,000 wind turbines have been installed from Shanghai to Texas, with continued expansion planned in coastal waters and on shore.

SEE ALSO: The Age of Sail (1571), Coal, $CO_2$, and the Climate (1896)

**Inventor Charles Brush** constructed this wind dynamo in the backyard of his Cleveland mansion in the winter of 1887–88.

# THE GREAT WHITE HURRICANE

O N THE AFTERNOON OF MARCH 11, 1888, a light snow began to fall over the northeastern United States. By the following morning, there were 18 inches (45.7 centimeters) of snow on the ground. But that was just the beginning. By midnight, 33 inches (83.8 centimeters) of snow had accumulated. And still it fell. When the storm that would end up being called the Great White Hurricane swept past Nova Scotia on March 14, some 40 to 50 inches (101.6 to 127 centimeters) of snow paralyzed parts of Connecticut, Massachusetts, New Jersey, and New York. Some 200 ships went down from the Chesapeake Bay to Maine.

While the storm was known by many as the Great White Hurricane, it was, in fact, a blizzard, a term first used to describe a snowstorm in the 1870s. The U.S. National Weather Service defines a blizzard as a storm with large amounts of snow and/or winds over 35 miles (56 kilometers) per hour, with visibility of less than a quarter mile (.40 kilometer), lasting for more than three hours. Blizzard conditions tend to develop on the northwest side of a storm when low pressure at the center and the higher pressure to the west intensify the circulation. Strengthening winds can pick up and blow snow around, creating whiteouts and significant snowdrifts.

In 1888, the gale-force winds produced drifts of more than 50 feet (15.2 meters) in metropolitan areas like New York City. (The storm would help convince city officials to build a subway system.) The city of Albany, New York, was shut down, and with no coal deliveries possible, thousands of residents were left without heat. Doctors couldn't make house calls because country roads were impassable. In the end, more than four hundred people died in the blizzard of 1888, half of them in New York City, making it the worst death toll from a winter storm in U.S. history.

SEE ALSO: The Great Blue Norther (1911), Fastest Wind Gust (1934)

**The New York City blizzard of 1888** is still considered to be the worst ever experienced, in large part because modern technology has since eased the hardships of extreme snowfall.

# DEADLIEST HAILSTORM

HAIL CAN EXACT TERRIBLE DAMAGE, causing billions of dollars in destruction to property and sometimes even killing people who can't find shelter. The frozen knobbly balls, formed as wind-borne precipitation in powerful storms freezes and refreezes, can occasionally reach softball size and weigh up to two pounds (907 grams).

The worst death counts in modern history were in India on April 30, 1888, when 246 people were killed by hail, and in Bangladesh on April 14, 1986, when hail reported to be the size of grapefruit killed ninety-two people. In 2017, the World Meteorological Organization determined that the 1888 event was the deadliest hailstorm on record.

But there are hints that a mysterious disaster in the Himalayas may have been much worse, with as many as 600 victims. Indeed, for the moment at least, Guinness World Records still gives the title for deadliest hailstorm to this incident, dated 850 CE.

The chilling scene of this meteorological disaster was discovered in 1942, when a British park ranger was probing Roopkund Lake, a tiny blue body of water surrounded by frozen gravelly slopes and fed by meltwater from glaciers. Skulls, bones, and other human remains could be seen in the clear water. At first, it being World War II, the theory was that a Japanese military unit had perished trying to cross the mountains. But it soon became clear the remains were ancient, preserved by the cold.

The cause was unresolved until 2004, when a team of forensic scientists sent by the National Geographic Channel dated the remains to the year 850 CE and found, hauntingly, that the deaths all seemed to be from severe blunt blows to the head and shoulders. They saw no other plausible cause than hail.

**SEE ALSO:** The Great Blue Norther (1911), Dangerous Downbursts Revealed (1975)

**Preserved human bones** can be seen in Roopkund, a glacial lake in the Indian Himalayas that was the scene of a deadly hailstorm in 850 CE, and is known by the grisly name of "Skeleton Lake."

# FIRST INTERNATIONAL CLOUD ATLAS

THE CLASSIFICATION OF CLOUDS PROPOSED in an 1802 lecture by an observant pharmacist, Luke Howard, was only the first attempt at what seems to have become a never-ending effort by meteorologists and amateur skywatchers alike to identify and understand distinctive patterns in these ephemeral features of the atmosphere. By the late 1800s, several atlases of clouds had been produced, and the precursor to the World Meteorological Organization sought to coordinate the publication of a standardized reference.

A "Clouds Commission" sifted through terminology, imagery, and methods to settle on a standardized set of descriptions that all scientists could utilize and employ in training students. In 1896, the first *International Cloud Atlas* was published, featuring color photographs—then, a rarity—showing a variety of familiar cloud types such as cirrus and cumulus along with more dramatic forms like the knobbly mamato-cumulus clouds found beneath some severe thunderstorms. The effort was facilitated by innovative photographic methods designed to increase contrast—including capturing the image of a cloud reflected in a quiet lake surface or a dark mirror.

Every decade or two, updated editions have been published, reflecting ever more varieties of cloud forms and including more sophisticated imagery. As in 1802, when Luke Howard stirred the emerging field of meteorology with his cloud classification, amateurs are still providing insights to professional meteorologists. The editors of the 2017 edition of the atlas added a new cloud type, the asperitas cloud, after reviewing suggestions and images from the Cloud Appreciation Society in Great Britain and a cloud-spotting society in Cedar Rapids, Iowa. The atlas is now online at wmocloudatlas.org, where you can explore an extraordinary gallery including images of that wavy, wispy asperitas cloud along with some other new arrivals, including the cavum, also known as the "hole punch" cloud.

**SEE ALSO:** Luke Howard Names the Clouds (1802), Beaufort Classifies the Winds (1806)

**In 2017, the cavum cloud, also called a fallstreak hole or "hole punch" cloud,** was among a batch of new cloud forms added to the International Cloud Atlas. These features typically appear in clouds formed of super-cooled water droplets. Most are circular, but elongated fallstreak clouds can be formed by passing aircraft. This one was photographed over Ann Arbor, Michigan, in November 2016.

# COAL, CO₂, AND THE CLIMATE

AS THE NINETEENTH CENTURY CAME TO an end, the connections between carbon dioxide, climate, and the increasing use of coal (and eventually other fossil fuels) began to emerge through the work of Svante Arrhenius (1859–1927), a Swedish chemist who was fascinated with climate and geology. The first step came in his remarkable treatise, "On the Influence of Carbonic Acid in the Air upon the Temperature of the Ground," published in April 1896. (Carbonic acid was the common term at the time for carbon dioxide.) As the historian James Rodger Fleming has described, Arrhenius drew together other scientists' findings in a model for the energy flows shaping climate, including processes that could amplify or counteract the warming influence from carbon dioxide and water vapor. Arrhenius roposed how a substantial rise or fall in the concentration of carbon dioxide in the air could explain the extent of ice ages and warm intervals.

In the prescient 1908 book *Worlds in the Making*, Arrhenius wrote that natural processes removing carbon dioxide from the air were outpaced by the increase from burning fossil fuels. "[T]he percentage of carbonic acid in the air must be increasing at a constant rate as long as the consumption of coal, petroleum, etc., is maintained at its present figure, and at a still more rapid rate if this consumption should continue to increase as it does now," he wrote.

Assuming the resulting rise in CO₂ and temperature would be gradual, he foresaw benefits, not risks, writing, "By the influence of the increasing percentage of carbonic acid in the atmosphere, we may hope to enjoy ages with more equable and better climates, especially as regards the colder regions of the earth, ages when the earth will bring forth much more abundant crops than at present, for the benefit of rapidly propagating mankind."

From the global warming calculations of English engineer Guy Callendar (1897–1964) in the 1930s, through those of Canadian physicist Gilbert Plass (1920–2004) in the 1950s, more observations and insights solidified the basic science, but led to a more unsettling forecast.

**SEE ALSO:** Scientists Discover Greenhouse Gases (1856), The Rising Curve of CO₂ (1958), Climate Models Come of Age (1967)

**Svante Arrhenius**, the Swedish scientist who studied the effects of carbon dioxide on the climate, pictured here c. 1920.

# A MIGHTY STORM

ON SEPTEMBER 8, 1900, IN THE DAYS before hurricanes had names, a great storm devastated Galveston, Texas. It remains the worst natural disaster in the history of the United States, with estimates of 6,000 to 12,000 people killed. Until then, Galveston had been a pearl of the Gulf Coast, a wealthy city that considered itself a rival to New Orleans. The great Sarah Bernhardt had sung in its opera house, and the state's first telephone and first electric lights were installed there. It was, until that September day, a city of the future.

The horrors of the storm, which made landfall as what is now known as a Category 4 hurricane, inspired the haunting song, "Wasn't That a Mighty Storm." First recorded by the preacher "Sin-Killer" Griffin, the song tells us that "Death come a-howlin' on the ocean, And when death calls, you got to go."

In the aftermath of the storm, Galveston built a seawall 17 feet (5.2 meters) high, which would eventually stretch more than 10 miles (16 kilometers). In an even greater feat of civil engineering, the "grade raising" workers jacked up Galveston's buildings by several feet (some by more than 10 feet, or 3 meters) and loaded dredged soil underneath. The rebuilding took more than ten years, but Galveston did not return to its old prosperity. Houston, which by 1914 had dredged its deep-water ship channel, grabbed much of the business and industry that Galveston had lost.

Unfortunately, the Gulf Coast would face more death and destruction at the hands of devastating storms. In 2005, New Orleans lost more than 1,245 citizens when floodwalls and levees couldn't keep out the surge from Hurricane Katrina. The costliest natural disaster in the United States, Hurricane Katrina is estimated to have caused $75 billion in damages in the New Orleans area and along the Mississippi coast, according to the U.S. National Hurricane Center. Residents of other cities along the Gulf Coast know with odds raised by rising seas and climate change, they too could someday face such a mighty storm.

*—J. S.*

SEE ALSO: The Great White Hurricane (1888), "China's Sorrow" (1931), The Human Factor in Weather Disasters (2006)

**A woman makes her way through the debris** on 19th Street in Galveston, Texas, in the aftermath of the 1900 hurricane that devastated the city.

# "MANUFACTURED WEATHER"

FROM THE EARLIEST TIMES, PEOPLE HAVE sought ways to keep cool when it's hot, allowing societies to develop in scorching desert climates and steaming jungles. In ancient Rome, cooling water from aqueducts circulated within the walls of the wealthy. One emperor ordered tons of snow brought by donkey cart to his gardens. Around 200 CE, a Chinese engineer built a rotating fan system for a palace, powered by slaves turning a crank.

In the mid-1800s, John Gorrie (1803–55), a Florida physician, envisioned cooling cities to relieve residents of "the evils of high temperatures." He started with hospitals, devising a rudimentary ice-based system for cooling patients' rooms. But half a century would pass before a young engineer, Willis Carrier (1876–1950), set the stage for widespread affordable air-conditioning—one of the biggest single changes in humanity's relationship with weather.

While working for the Buffalo Forge Company in 1902, Carrier was asked to solve a humidity problem at a printing company. Through a series of experiments, he designed a system that controlled temperature and humidity using coolant-filled coils. He then secured a patent for his "Apparatus for Treating Air." Around the same time, a ventilation expert, Alfred R. Wolff (1859–1909), designed the first systems specifically built to cool crowded rooms—with the first unit installed in 1903 at the New York Stock Exchange.

Carrier and some partners launched a company in 1915 bearing his name and using the slogan, "Manufactured Weather for any requirement." By the 1920s, air-conditioning's use was fast spreading from printers and candy makers to "comfort cooling" for movie theaters and department stores.

The invention of affordable individual home cooling units helped spur the spread of suburbia in the steamy south, and air-conditioning is now a key to the growth of cities worldwide. But the boom is creating fresh challenges. Some refrigerants can add to global warming if they escape, and if the additional electricity required to power the units is generated with fossil fuels, that can worsen pollution and produce more greenhouse gases.

SEE ALSO: An Umbrella for Everyone (1830), Settling a Hot Debate (2012)

**A woman in the 1960s** poses with her air conditioner, a comfort that first become available early in the twentieth century and dramatically altered people's relationship with weather.

M. ANDERSON.
WINDOW CLEANING DEVICE.
APPLICATION FILED JUNE 18, 1903.

NO MODEL.

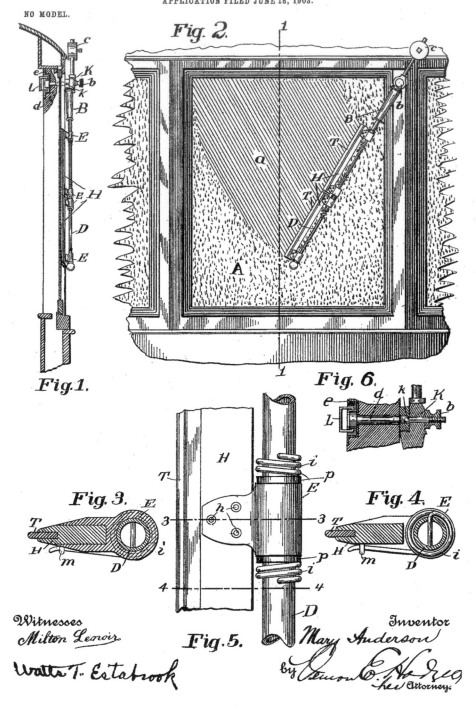

Fig. 2.

Fig. 1.

Fig. 6.

Fig. 3.          Fig. 5.          Fig. 4.

Witnesses
Milton Lenoir
Watts T. Estabrook

Inventor
Mary Anderson
by _____ C. Hedge,
her Attorney.

# THE WINDSHIELD WIPER

TODAY, THE WINDSHIELD WIPERS THAT maintain a clear view in a speeding vehicle on a stormy day are as unremarkable as shoelaces or a toothbrush. They are just there. But the first workable version of this device was conceived in a trolley car in New York City on a winter day in 1902, by Mary Anderson (1866–1953), a visiting real-estate developer from Alabama. She had noticed that the motorman had to drive the trolley with both panes of the windshield open because he couldn't clear the glass of accumulating sleet.

According to various accounts, she began sketching solutions on the spot, and then, back home in Alabama, pursued mechanical models with the help of a designer and a local company. On November 10, 1903, Anderson was granted U.S. Patent No. 743801, which includes her succinct description:

> [I]t will be seen that a simple mechanism is provided for removing snow, rain, and sleet from the glass in front of the motorman, and it is simply necessary for him to take hold of the handle L and turn it in one direction or the other to clean the pane, the spring action upon the cleaners operating to hold the rubbers in yielding contact against the glass with sufficient pressure to clean the latter and at the same time with sufficient yielding action so as not to be rendered inoperative by striking an obstruction. In this way the difficulty of not being able to see through the front glass in stormy weather is effectually obviated.

Anderson's efforts to sell the invention came to naught and the patent expired seventeen years later, before the auto industry had its revolutionary boom. Other versions had been invented and, in 1922, Cadillac became the first automobile manufacturer to offer wipers as standard equipment.

In 2011, Anderson was inducted into the National Inventors Hall of Fame.

**SEE ALSO:** An Umbrella for Everyone (1830)

**Diagram of Mary Anderson's** 1903 patented window cleaner design.

# A DRY DISCOVERY

WHEN ONE THINKS OF DRY REGIONS, it is hard to avoid iconic visions of shimmering, sun-scorched sands. Indeed, lists of the world's driest spots are dominated by such places—from the deserts around Egypt's Aswan Dam, to Chile's Atacama Desert, the second-driest place on Earth, where some spots, shielded from west and east by mountain ranges, are thought to have been rain-free for 500 years. Indeed, NASA scientists have used the Atacama Plateau as a stand-in for Mars in some studies.

But the absolute parched stand-out is, of all places, in Antarctica—a continent dominated by a mile-thick sheath of water, albeit frozen. There, the McMurdo Dry Valleys are almost entirely bereft of moisture, with the flow of ice sheets from the interior blocked by high mountains and atmospheric humidity lost as cold, dense air from the high polar plateau flows down slopes from the interior, heating as it descends in gusts that sometimes hit hurricane speed.

The valleys were first sighted on December 18, 1903, on the first Antarctic expedition led by Robert Falcon Scott (1868–1912), the British Royal Navy officer and explorer who reached the South Pole with four men on a second expedition, in 1912, but perished with the others on the attempted return to the continent's coast.

On that first trip, Scott explored one of the valleys briefly, but did not linger. Scott later wrote: "[W]e have seen no living thing, not even a moss or a lichen; all that we did find, far inland amongst the moraine heaps, was the skeleton of a Weddell seal, and how that came there is beyond guessing. It is certainly a valley of the dead; even the great glacier which once pushed through it has withered away."

The Dry Valleys were more thoroughly surveyed by another team from Scott's second expedition and have been intensively studied in recent decades and, despite the harsh climate, found to harbor a host of "extremophiles," organisms capable of enduring conditions hostile to most life.

SEE ALSO: The Dust Bowl (1935), The Coldest Place on Earth (1983)

**The Wright Valley in Victoria Land**, Antarctica, was once occupied by a glacier but is now free of ice—and precipitation. The valley was named for Sir Charles Wright, a member of the 1910 British Antarctic Expedition.

# THE GREAT BLUE NORTHER

IN TEMPERATE REGIONS AROUND THE WORLD, powerful cold fronts occasionally produce dramatic flips in conditions from balmy to brutal. But it's hard to find a record of one as broad and potent as the system that swept the American heartland on November 11, 1911, and gained the name the Great Blue Norther.

In a recent analysis of this meteorological milestone, the U.S. National Weather Service said the stage was set two days beforehand when a broad high-pressure system built over Alberta, Canada. As a powerful low-pressure system pushed from the Rockies east into Iowa and Missouri, it drew unseasonably warm air northward ahead of it and cold northern air flowed in behind.

In Columbia, Missouri, the Blue Norther hit around 2 p.m and changed warm breezes to howling northerly gales, according to a University of Missouri study marking the centennial. In just one hour, the temperature dropped from 82 to 38 degrees Fahrenheit (27.8 to 3.3 degrees Celsuis) and by midnight had plunged to 13 (-10.6°C).

A weather dispatch from Springfield, Missouri, reported that around 2:30 p.m., "a dense greenish black bank of clouds was rising along the western horizon." The same extraordinary pattern was seen in dozens of communities across the region.

Potent thunderstorms and tornadoes tore through towns from the Mississippi Valley up across the states around the Great Lakes, killing more than a dozen people.

In Chicago, one man was overcome by heat and two others froze to death within a twenty-four-hour period, the Weather Service recounted.

The system remained destructive even as it moved across the East Coast, with fourteen crew members reportedly perishing after a wind-tossed barge broke free from a tugboat off the New England coast.

SEE ALSO: The Great White Hurricane (1888), Fastest Wind Gust (1934)

**On November 11, 1911,** a tornado spawned by an extraordinary cold front devastated Owosso, Michigan, including this furniture factory.

# ORBITS AND ICE AGES

STARTING IN THE MID-NINETEENTH CENTURY, the emerging understanding of great past cycles of ice ages and warm intervals, including the present one, prompted a hunt for explanations. Among the first scholars to point the discussion away from Earth—to subtle variations in its orbit and its orientation to the sun—was James Croll (1821–90), a remarkable man who taught himself physics and astronomy by borrowing books while working as a janitor at a university museum in Glasgow.

Croll ended up corresponding with a leading analyst of ice ages, Charles Lyell, and that led to a position at the Geological Survey of Scotland. In 1875, his calculations and concepts were laid out in a book with a title that nicely captured the question of the day: *Climate and Time, in Their Geological Relations*. He calculated that in some periods, varying on scales of tens of thousands of years, the Northern Hemisphere would get slightly less sunlight, allowing snow to build and leading to an ice age. The notion was strongly resisted.

The next scholar to explore this relationship was Milutin Milanković (1879–1958), a Serbian engineer and mathematician who became fascinated with astronomy and climate history in the early twentieth century. Beginning in 1912, for more than a decade he laid out mathematically how three variations in Earth's orbit and orientation toward the sun could periodically produce summers with unmelting snows, creating dynamics that could build the great ice sheets. His keystone thesis, "Climates of the geological past," was published in 1924—and followed by more decades of debate.

Evidence began to accrue as scientists developed methods for dating terrestrial materials and fossilized ocean plankton using variations in isotopes of carbon and oxygen. While many questions remain unresolved, Milanković's theory was strongly buttressed when a landmark paper in 1976 confirmed that the basic cycles were reflected in layered sediment extracted from seabeds.

SEE ALSO: Ice Ages Revealed (1840), Space Weather Comes to Earth (1859), Climate Clues in Ice and Mud (1993)

**A photograph of the sun over Earth's horizon**, taken from the International Space Station in 2013. From the late 1800s through the time of the pivotal work of Milutin Milanković starting in 1912, scientists built the case that ice ages were shaped by subtle changes in Earth's orientation toward the sun.

# A "FORECAST FACTORY"

THE IDEA OF PREDICTING THE WEATHER BY solving mathematical equations that represent atmospheric conditions was formulated in 1904 by the Norwegian meteorologist Vilhelm Bjerknes (1862–1951). The first attempts to move from theory to practice were made by Lewis Fry Richardson (1881–1953), a British mathematician with a penchant for putting science to practical use. For instance, horrified by the 1912 sinking of the *Titanic*, he demonstrated and gained a patent for a way to use horn blasts from a ship to determine the location of icebergs.

His ideas for numerical weather forecasts emerged a few decades before computers provided the number-crunching power necessary for such calculations, but they form the intellectual underpinning for the basics of weather and climate modeling since.

In 1916, Richardson began the first test of his method, which involved representing atmospheric dynamics in three dimensions with a set of equations. His task was to generate a six-hour "forecast" for central Europe starting at 7:00 a.m. on May 20—of 1910! A critical step in weather prediction is knowing initial conditions well enough to estimate how things might change as time passes. He chose this date because he had abundant data on humidity, barometric pressure, and winds recorded at the time by Bjerknes. In a heroic effort, much of it undertaken in spare moments while he served as an ambulance driver during the First World War, he spent the next two years performing the grueling calculations to create that forecast.

And the forecast, in the end, was a failure, as he gamely explained in his remarkable 1922 book *Weather Prediction by Numerical Process*. But he was convinced the method was sound. In the book, Richardson envisioned a "forecast factory" using human "computers." Directed by a central coordinator, 64,000 people in a massive arena would each do the calculations for their allocated part of the globe.

The first electronic, computer-calculated weather forecast was produced in 1950 by Jule Charney (1917–81) and others at the Aberdeen Proving Ground, in Aberdeen, Maryland. Richardson was thrilled by the news, calling it an "enormous scientific advance."

—*P. D. W.*

SEE ALSO: First Weather Forecasts (1861), The First Computerized Forecast (1950), Climate Models Come of Age (1967)

**The artist Stephen Conlin** created this view of Lewis Fry Richardson's proposed weather "forecast factory in 1986.

# "CHINA'S SORROW"

MANY GREAT RIVERS HAVE BEEN A source of both bounty and peril, providing fertile soil, trade routes, and abundant water but also delivering devastating floods. This pattern is nowhere more pronounced than along the meandering course of China's Yellow River, or Huang He. Historians describe the river as the cradle of Chinese civilization but have also called it "China's Sorrow" because of the extraordinary death counts when rising waters have periodically breached its silty banks.

The river had flooded more than a thousand times in four thousand years. But a handful of particularly devastating floods are estimated to have taken a million or more lives, most notably in 1887, when between 900,000 and 2 million people died. Then, in 1931, 34,000 square miles (88.060 square kilometers) were inundated, leaving tens of millions homeless. The rising waters and subsequent disease and famine are variously estimated to have killed between 850,000 and 4 million people. That year also saw terrible flooding along the Yangtze and Huai Rivers.

Heavy rains were a factor in most flood years, but it would be a distraction to focus on meteorology in probing why the Yellow River is the deadliest river in the world. The level of danger has mainly been a function of changes in population and land use and centuries of efforts to control the river's course on its way to the sea—fighting powerful dynamics created by the massive amount of transported sediment that gives the river its name.

Most of that sediment is carved out of the deep layers of fine yellow dust that accumulated over millions of years on the Tibetan Plateau, much of it carried on desert winds during a cool, dry period between 7 and 8 million years ago. Downstream, over time as the sediment has been deposited along the river's course, the main channel has ended up elevated above surrounding plains—in some places up to 30 feet (9 meters) higher. In a pattern peaking some three centuries ago, work to rein in the river simply worsened the long-term risk.

SEE ALSO: A Mighty Storm (1900), North Sea Flood (1953)

**A 2008 photograph of the Hukou Waterfall** along the surging, and occasionally devastating, Yellow River in China.

# THE FASTEST WIND GUST

1934

WHILE NEW HAMPSHIRE'S MOUNT Washington, at 6,288 feet (1,917 meters), is the tallest mountain in the northeastern United States, it is still 22,741 feet (6,931.5 kilometers) shorter than Mount Everest. Yet winter wind and weather conditions on the American peak rival those on the Himalayan giant.

Its location puts Mount Washington at the confluence of several major tracks for North American storm systems. The jet stream carries storms west to east over the mountain, while the north-to-south orientation of the state's Presidential Range makes it a barrier to these westerly winds. The winds from the jet stream intersect with weather systems moving south to north along the coast. Additionally, Mount Washington is situated at the throat of a funnel so that winds from the northeast are channeled toward it. The steep western face of the mountain compresses wind caught in the funnel even more.

All of these factors combine to make Mount Washington one of the windiest places on Earth. On average, hurricane wind gusts are observed at its summit 110 days a year.

For nearly sixty-two years, the observatory atop the peak—often frosted in ice and snow—held the record for the fastest wind gust measured on the planet: a 231-mile-per-hour (372 kilometers-per-hour) blast measured by observatory staff on April 12, 1934. This record was broken in 1996 when an unmanned weather station on Barrow Island, Australia, recorded a wind gust of 253 miles per hour (407 kilometers per hour) during Typhoon Olivia.

The wind gust on Mount Washington, however, remains the highest surface wind speed ever observed directly by people. On that day, the staff at the Mount Washington Observatory, including Salvatore Pagliuca, Alex McKenzie, and Wendell Stephenson, woke to what they recorded as a "super hurricane, Mount Washington style." As the morning wore on, the winds became stronger and stronger. At 1:21 p.m., the anemometer recorded the gust of 231 miles (372 kilometers) per hour. After this wind measurement, the National Weather Bureau ran the anemometer through a number of tests confirming that the measurement was valid.

**SEE ALSO:** The Great White Hurricane (1888), The Great Blue Norther (1911)

**The Mount Washington Observatory** in New Hampshire, which experiences some of the world's most extreme weather, is routinely coated in rime ice formed by freezing wind-borne water droplets.

# THE DUST BOWL

OVER THOUSANDS OF YEARS, RICH SOIL accumulated beneath deep-rooted grasses across America's Great Plains to an extraordinary depth of 6 feet (1.8 meters). In the late nineteenth century, waves of pioneers settled to graze cattle and raise crops. Into the 1920s, generous federal incentives and rising wheat prices led to the "Great Plow-Up" in the southern plains, with more than 5 million acres of a grassland ecosystem that had evolved to withstand drought replaced with tilled fields. As the Great Depression hit, wheat prices plunged and lands were abandoned.

Starting in the summer of 1930, a severe drought developed and the price of decades of destructive farming practices came due. On May 9, 1934, a towering wall of dust 10,000 feet (3,048 meters) high rolled across the plains, thick enough to block out the sun. The storm gathered strength as it traveled east carrying 350 million tons (318 million tonnes) of soil in huge clouds. The dust storm blew all the way to Chicago, where it dumped around 12 million pounds (5.4 million kilograms) of soil on the city. Two days later, the storm reached New York City where the sky darkened and a dark blanket settled. The storm continued to Boston and then out to sea. Crews on ships in the Atlantic Ocean found themselves having to sweep decks clear of a thick layer of Great Plains soil.

That was just the beginning. The worst dust storms occurred on April 14, 1935, which became memorialized as "Black Sunday." Dust storms swept across the Great Plains from Canada south to Texas, causing extensive damage and turning day to night. Hundreds of people suffered and many died from "dust pneumonia." Robert E. Geiger, a Denver-based reporter who happened to be in Oklahoma that day, was the first to use the term *Dust Bowl*. The Dust Bowl added greatly to the misery already besetting the heartland in the midst of the Great Depression.

SEE ALSO: Long-Distance Dust (2006), Settling a Hot Debate (2012)

**A potent dust storm** approaching Stratford, Texas, on April 18, 1935.

# Le Petit Journal

**ADMINISTRATION**
61, RUE LAFAYETTE, 61

Les manuscrits ne sont pas rendus

On s'abonne sans frais
dans tous les bureaux de poste

5 CENT.

27me Année

SUPPLÉMENT ILLUSTRÉ

Numéro 1.307

**DIMANCHE 9 JANVIER 1916**

5 CENT.

**ABONNEMENTS**

| | SIX MOIS | UN AN |
|---|---|---|
| SEINE et SEINE-ET-OISE | 2 fr. | 8 fr. 50 |
| DÉPARTEMENTS | 2 fr. | 4 fr. |
| ÉTRANGER | 2 50 | 5 fr. |

# LE GÉNÉRAL HIVER

# RUSSIA'S "GENERAL WINTER"

WEATHER HAS OFTEN PLAYED AN unpredictable role in the outcome of wars, as when a change in the winds helped Britain's fleet defeat the more powerful Spanish Armada in 1588. But sometimes its importance, even when highly predictable, has been underappreciated. This has been especially true when it comes to invasions of Russia, so famed for notorious cold and paralyzing soggy thaws that historians of war have written of "General Winter" and "General Mud" as battlefield foes in that country.

Whether in Sweden's failed 1708 invasion during the Great Northern War or Napoleon's try in 1812, the cold was generally not the only, or even the decisive, factor. But it was always there, killing, crippling, and debilitating troops. In Germany's attempt to crush Russia in 1941, Hitler's overconfidence led to a delayed approach to Moscow, allowing winter to join the fray.

In his 2011 book, *The Storm of War*, historian Andrew Roberts recalled how, on December 20, 1941, Minister of Propaganda Joseph Goebbels appealed to German citizens for warm clothing to send to the front: "Those at home will not deserve a single peaceful hour if even one soldier is exposed to the rigors of winter without adequate clothing."

It was too little, too late.

Hitler's dismissive attitude toward his weather forecasters could well have contributed to the disastrous setback. In a monologue on meteorology recorded late on the night of October 14, 1941, he made his views clear:

> One can't put any trust in the [meteorological service] forecasts. . . . Weather prediction is not a science that can be learnt mechanically. What we need are men gifted with a sixth sense, who live in nature and with nature—whether or not they know anything about isotherms and isobars. . . .

In his account, Roberts noted that Hitler's library contained many books on Napoleon's campaigns. With some irony, he added, "Yet he did not learn the most obvious lesson from his predecessor."

SEE ALSO: The Age of Sail (1571), The Jet Stream Becomes a Weapon (1944)

**An illustration of "General Winter,"** Russia's frequent wartime ally, on the eastern front of World War I, as seen on the cover of the French newspaper *Le Petit Journal*, January 9, 1916. From Napoleon through Hitler, Russia's enemies faced this tough opponent.

MARCH 1950

35 CENTS

# POPULAR
# MECHANICS
## MAGAZINE
WRITTEN SO YOU CAN UNDERSTAND IT

REG'D. TRADE MARK, GREAT BRITAIN, No. 410426

REG. U.S. PAT. OFF.

## FLY INTO THE HEART OF A TYPHOON
*—Read this terrific story of a B-29 crew—Page 133*

# HURRICANE HUNTERS

IT IS NOW COMMON FOR SPECIAL AIRCRAFT TO fly into the heart of tropical storm systems of all strengths to collect data that can improve U.S. National Hurricane Center forecasts. The flights produce details on wind speed and other conditions that satellite images or radar cannot detect. Critical to the effort are dropsondes, instrument-packed tubes dropped into a storm that transmit a range of atmospheric conditions as they descend. Up to 1,500 are deployed during such missions in a typical hurricane season.

Such flights date back to World War II, with military aircraft patrolling the Pacific long before the era of weather satellites, to try to keep track of typhoons. The first official hurricane flights began in 1944, but the first known flight into a hurricane's eye took place one year before—on a dare and a wager.

In 1943, a group of British pilots were stationed at Bryan Field in Texas while being trained in flying by instruments alone—a new set of protocols and practices designed for safe flying at night or in bad weather. The instructor was one of the early masters of this method, Air Force Colonel Joe Duckworth (1902–64).

Word came that a hurricane, later referred to as the "Surprise" Hurricane of 1943, was intensifying and approaching Galveston—which had been so devastated by the great hurricane of 1900.

When the British airmen heard that the two-seat AT-6 "Texan" aircraft they were training in might have to be flown to a safer location, they ribbed their instructors about what they perceived as a sign that the planes were frail. According to an account of the incident written by Lew Fincher, a meteorologist and historian, Duckworth proved them wrong, on a bet, by flying into the storm with a navigator from the base, returning safely and repeating the feat.

Flying into hurricanes quickly became a serious scientific enterprise, conducted with four-engine aircraft. It is also a dangerous occupation. Six hurricane-hunter aircraft were lost in storms between 1945 and 1974—five in the Pacific and one in the Caribbean—taking fifty-three crew members with them.

SEE ALSO: Franklin Chases a Whirlwind (1755), First Photographs of Tornadoes (1884), A Mighty Storm (1900), Storm Chasing Gets Scientific (1973)

The March 1950 issue of *Popular Mechanics* told the story of a B-29 bomber used by the U.S. Air Force to probe typhoons in the Pacific.

# THE JET STREAM BECOMES A WEAPON

THE TERM *JET STREAM* ENTERED METEOR-ological literature in a 1947 paper by University of Chicago scientists describing meandering high-altitude, high-velocity rivers of air. Attention had been drawn to such winds as Allied bombers in World War II found their westward progress surprisingly slow. But these winds had been discovered two decades earlier—in work that would actually play an unexpected, and deadly, role during the war.

Between 1923 and 1925, a Japanese meteorologist, Wasaburo Ōishi, launched more than 1,200 small balloons from an observation station north of Tokyo to estimate wind speeds at different altitudes in different seasons. In winter, he identified a layer of winds that often exceeded 150 miles (241 kilometers) per hour, blowing from the west at an altitude between 25,000 and 35,000 feet (7,620 and 10,668 meters). He published his findings in 1926, but just in the observatory's journal and written in Esperanto—an attempt at a universal language.

Ōishi passed away in 1940 but his insights survived. Guided by his charts, Japan launched 9,000 balloons laden with incendiary bombs between November 1944 and April 1945. Most fell into the Pacific. But 300 made it to the United States.

Nearly all of those that did arrive landed in unpopulated areas. But one of the fire-bombs did claim victims. On a beautiful May afternoon in Oregon in 1945, a pastor, Archie Mitchell, and his pregnant wife, Elyse, took five children from their Sunday school class on a drive to enjoy a picnic in the woods. While Mitchell sought advice from a crew doing road work, Elyse and the children explored the forest nearby. Richard Barnhouse, one of the road-crew workers, saw Elyse and the children pointing at something on the ground. Then there was a large explosion. A fiery blast killed Elyse and all of the children—the only casualties from enemy action on the North American continent, killed by a bomb carried across the ocean on a paper balloon.

SEE ALSO: First Weather Balloon Flight (1783), Hurricane Hunters (1943)

**Late in World War II, Japan exploited the jet stream** to loft thousands of balloons carrying incendiary bombs toward the United States. About 300 are known to have reached the mainland. A pregnant woman and five children died when they encountered one in southern Oregon on May 5, 1945.

# RAINMAKERS

PEOPLE HAVE DREAMED OF CONTROLLING the weather through much of history. Most attempts relied on magic or prayer until the nineteenth century, when a few flamboyant characters pursued more physical, if still rather fantastical, approaches. As documented in detail by James Rodger Fleming, a Colby College historian of science and technology, a string of men garnered headlines as "rainmakers" for trying to trigger precipitation or otherwise tame the elements. In the 1830s, James Espy (1785–1860), who later became the first meteorologist employed by the United States federal government, proposed setting massive fires to stimulate storms by creating upwellings of hot air. In 1891, Congress lavishly funded a bizarre effort using explosions to break a Texas drought. And an entrepreneur with a secret mix of rain-making chemicals, Charles Hatfield (1875–1958), brought rainmaking, or "pluviculture," into the twentieth century when he was hired by several municipalities to end dry spells by mysteriously evaporating his potion from wooden platforms. A deadly California flood coincided with one effort, landing him in a lawsuit.

A more scientific approach appeared to bear fruit in 1946, when researchers at General Electric's laboratory in Upstate New York came up with ways to seed clouds—a critical step on the path toward making rain or snow. Vincent Schaefer (1906–93), a lab technician, found that dry ice instantly transformed moisture in a cloud chamber into millions of ice crystals. Another researcher, Bernard Vonnegut (1914–97), a brother of the novelist Kurt Vonnegut, soon discovered that silver iodide particles had a similar effect. In November 1946, Schaefer and a pilot dispersed silver iodide from a small plane high over Massachusetts, producing a veil of snow.

Cloud-seeding quickly moved from tests into applications, including in warfare, resulting in a 1978 treaty barring military modification of the environment. While scientific studies have yet to demonstrate a significant effect, it is still attempted regularly in China, parts of the United States, and the Middle East.

SEE ALSO: "Manufactured Weather" (1902), Climate by Design? (2006)

**In 1946, Vincent Schaefer,** a scientist at General Electric, devised methods for producing clouds using dry ice. Silver iodide was also used and tested.

# THE FIRST COMPUTERIZED FORECAST

AS LONG AGO AS THE LATE NINETEENTH century, the American meteorologist Cleveland Abbe and peers recognized that the flows of energy and moisture in the atmosphere shaping weather could be characterized mathematically. The prospect of useful predictability was clear, but elusive. Lewis Fry Richardson's 1922 vision of a "forecast factory" created the architecture for numerical weather prediction. But another couple of decades would pass before two critical needs would be satisfied—sufficient flows of observations of real-time conditions across wide stretches of the planet and the massive capacity for rapid calculations that was only possible with computers.

The quality of a weather forecast lies not only in the design of a mathematical simulation of the atmosphere, but also in the ability to reproduce with as much detail as possible the initial conditions across landscapes and seascapes at a point in time—winds, temperature, pressure. Global connectivity and fast-expanding networks of weather stations were rapidly improving that picture. Satellites and deep-diving ocean buoys would soon enable vastly greater leaps.

But there was also brute number crunching. The impetus for progress on this aspect of weather forecasting, as in so many realms—from spaceflight to energy technology—came from its relevance to national defense. With the Cold War brewing, the navy came through with support for John von Neumann (1903–57), a heralded mathematician who led the Institute for Advanced Study in Princeton, New Jersey. The nation's Meteorological Research Project began in July 1946.

Von Neumann assembled a crack group of meteorologists, led by Jule Charney, to refine a mathematical model of the atmosphere. They had access to the first computer, the 30-ton, 18,000-tube ENIAC machine, which had first been used in 1945 to calculate the feasibility of a hydrogen bomb. The first twenty-four-hour computer weather forecast was completed in April 1950 and published later that year. It took more than twenty-four hours for the computer to complete the calculations, but the effort demonstrated the technique worked. Three years later, the Swedish BESK computer generated the first real-time numerical weather prediction, beating the actual weather by some ninety minutes.

**SEE ALSO:** First Weather Forecasts (1861), A "Forecast Factory" (1922)

**Marlyn Meltzer (standing) and Ruth Teitelbaum (crouching)** are seen programming the ENIAC computer for the U.S. Army in 1946. Many early computer programmers were women.

# TORNADO WARNINGS ADVANCE

A CENTRAL GOAL IN METEOROLOGY HAS long been to warn of severe weather. But there's always been conflict between the need to warn and the risk of false alarms, particularly for the most dangerous kinds of storms. In 1878, John Park Finley (1854–1943), a young officer in the U.S. Army Signal Corps with substantial training in meteorology, began a sustained and detailed study of tornado locations and related weather conditions. He summarized his research in an illustrated book, *Tornadoes: What They Are and How to Observe Them; with Practical Suggestions for the Protection of Life and Property*. Published in 1887 by the Insurance Monitor, the book included protocols for projecting tornado risks.

But that same year, the Signal Corps instituted a ban on the word *tornado* in its forecasts. Hurricane warnings were also officially forbidden. The policy was enforced with particular vigor by Willis Moore, chief of the U.S. Weather Bureau from 1895 to 1913. (Moore's antipathy for issuing hurricane warnings is thought to have played a role in the horrific loss of life in the great Galveston hurricane of 1900.)

This tornado warning ban loosened in 1938 but persisted in practice for another decade. On March 20, 1948, a powerful tornado destroyed dozens of aircraft at a military airfield in Oklahoma. Two Air Force meteorologists were immediately tasked with assessing the predictability of such storms and just five days later issued a tornado warning three hours ahead of another twister that would hit the base.

On July 2, 1950, Francis W. Reichelderfer, the chief of the Weather Bureau from from 1938 to 1963, ended the ban, writing: "Whenever the forecaster has a sound basis for predicting tornadoes, the forecast should include the prediction in as definite terms as the circumstances justify."

A big advance came on a stormy day in Illinois, April 9, 1953, when a radar system left over from World War II detected a "hook echo" reflecting tornadic activity deep in a thunderstorm. This telltale sign and the increasing use of radar led to more reliable warnings. But these potent storm systems can still exact awful damage, particularly if warnings are not heeded or shelters not near at hand.

SEE ALSO: A Mighty Storm (1900), The First Computerized Forecast (1950)

**While tornadoes can still be devastating** and deadly, a combination of improved warnings and more structures adding or including shelters has dramatically cut death rates. Here a concrete domed shelter survived the Moore, Oklahoma, tornado on May 20, 2013.

# LONDON'S GREAT SMOG

COAL WAS A HEATING FUEL IN LONDON AS long ago as the Middle Ages. By the nineteenth century, as populations grew, coal burning was frequently linked to episodes of noxious air, as Charles Dickens noted in the *Dictionary of London* in 1882: "Nothing could be more deleterious to the lungs and the air-passages than the wholesale inhalation of the foul air and floating carbon, which, combined, form a London fog."

But all previous pollution pulses paled beside what came to be known as the Great Smog, which enveloped the city for five days starting on December 5, 1952. Through an unusually prolonged cold snap, coal stoves worked overtime heating homes. Under normal circumstances, the smoke would have risen into the atmosphere and dispersed, but a high-pressure system over the city caused a temperature inversion that trapped the building plume along with humid air.

As Britain's meteorological agency, the Met Office, has explained, the pollution could have acted as a catalyst for fog, with water condensing on the tiny particles. The resulting mix of chemicals and water created acidic conditions that could have worsened skin and breathing problems.

A toxic fog enveloped and paralyzed the city. Black slime coated sidewalks and roads. Double-decker buses had to be guided through the streets by a conductor walking ahead wielding a light. In the end, the acidic fog sent 150,000 people to hospitals and was linked to as many as 12,000 deaths. The incident resulted in the passage of the UK's Clean Air Act of 1956 and subsequent updates banning smoky emissions.

In recent years, the industrializing cities of China and India have had their turn grappling with terrible smog episodes fueled by a twenty-first-century brew of pollutants from coal burning, car exhausts, and cooking fires.

**SEE ALSO:** London's Last Frost Fair (1814), Coal, $CO_2$, and the Climate (1896), The Dust Bowl (1935)

**A man guides a London bus** through thick, noxious fog.

# NORTH SEA FLOOD

THROUGHOUT THEIR HISTORY, THE PEOPLE of the Netherlands—literally, the "Lowlands"—have had to work hard to keep the North Sea at bay. Twenty percent of the densely populated nation lies below sea level and another half of its territory is no more than a meter (3.28 feet) above the tide. One function of early windmills was to pump water out of polders, which are diked and drained tracts of formerly marshy land.

Concerns grew early in the twentieth century that a big storm could threaten the nation with widespread inundation. A 1937 government report, noting warning signs including the deterioration of dikes facing the sea, proposed a major project involving building barriers across sea inlets and other steps to reduce reliance on sea-facing dikes. But various delays, and then World War II, slowed the progress of these projects. By 1953, only two river mouths had been protected.

Then disaster struck. On January 31, hurricane-force winds produced a tremendous storm over the North Sea. At the time of the storm, many dikes were not high enough, and were eroding and weakened where military structures were built into them during the war. Sea dikes breached in 150 places. The storm surge, which coincided with a high spring tide, flooded 350,000 acres (1400 square kilometers) overnight, killing 1,836 people. While coastal areas of England and Belgium also flooded, the Dutch were hit the hardest.

As the nation slowly recovered, the Dutch government created the Delta Works Committee to figure out how to strengthen coastal defenses to withstand an unheard-of standard—fending off a one-in-ten-thousand-year storm. The resulting Delta Plan included blocking estuaries, building dams and storm surge barriers, installing sluices and locks, and strengthening dikes. Construction began in 1958 and was officially completed in 1997.

In 2014, taking into account projections of rising seas from global warming, a new Delta Plan was approved, under which another $25 billion in flood defenses would be spent over thirty years.

SEE ALSO: California's Great Deluge (1862), A Mighty Storm (1900), "China's Sorrow" (1931)

**Residents of Kruiningen, the Netherlands**, survey devastation following the North Sea flood. Many towns were still buried in mud six months after the waters receded.

# THE RISING CURVE OF CO$_2$

THROUGH A CENTURY OF STUDY, IT HAD become clear that the trace gas carbon dioxide was heating the planet, and accumulating emissions from burning fossil fuels could substantially add to warming. But no one yet knew how fast the concentration of this gas in the atmosphere might rise. Was absorption of CO$_2$ by the oceans and forests keeping up?

In 1957 two scientists from the Scripps Institution of Oceanography, Roger Revelle (1909–91) and Hans E. Suess (1909–93), helped determine that the oceans actually had a limited capacity to store carbon dioxide. Just before their paper was sent for publication, Revelle added a line that has been heavily cited ever since: "Human beings are now carrying out a large scale geophysical experiment of a kind that could not have happened in the past nor be reproduced in the future."

The "experiment" is the continuing buildup of carbon dioxide in the atmosphere. Its outcome would have substantial consequences for humans and ecosystems for centuries to come.

But consistent measurements were needed to track what was actually happening with CO$_2$. In 1958, as part of a sweeping research initiative called the International Geophysical Year, a young chemist at Scripps, Charles David Keeling (1928–2005), installed an instrument that could continually measure carbon dioxide levels 11,000 feet (3,352.8 meters) up on Mauna Loa, a massive dormant volcano on the Big Island of Hawaii. Far from any contamination sources, the recording began. The readings showed an annual dip and rise, reflecting the surge of plant growth each spring and summer in the Northern Hemisphere. But by 1960, a longer-term rise in the CO$_2$ concentration was evident and has become ever clearer. Keeling's systematic measurement of carbon dioxide concentration over the ensuing decades, carried on by his son, Ralph Keeling, after he died in 2005, has produced an iconic graph now dubbed the Keeling Curve.

SEE ALSO: Scientists Discover Greenhouse Gases (1856), Coal, CO$_2$, and the Climate (1896)

**Charles David Keeling** of the Scripps Institution of Oceanography examines the running record he developed of carbon dioxide concentrations.

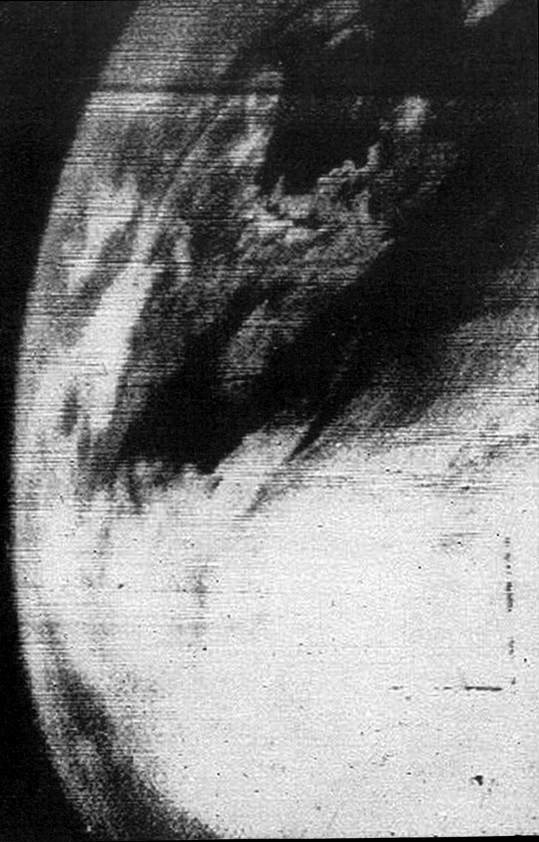

# WATCHING WEATHER FROM ORBIT

S THE 1950S ENDED, WEATHER FORE-casting was poised for a transformational leap. Computer simulation, radar, and other tools were producing rapid improvements, but the space race brought a new element—a view of weather patterns from Earth-focused cameras lofted into orbit. This was soon to be followed by instruments providing a host of readings, not only of cloud patterns but also of temperature, moisture, and much more.

The first attempt at launching a weather satellite came on February 17, 1959, when the United States launched *Vanguard 2*. But part of the rocket that delivered the satellite into orbit bumped into the spacecraft, producing a wobble as it orbited that greatly limited its collection of useful data on cloud cover.

On April 1, 1960, the era of satellite meteorology began in earnest with NASA's successful launch of *TIROS-1*, a 270-pound spacecraft equipped with two solar-powered television cameras, one capturing a wide view of Earth and the other a tighter image. The cameras snapped photos every thirty seconds as the satellite orbited from one pole to the other,

storing images on a magnetic tape recorder when the satellite was out of range of the ground station network.

The first image captured by the *TIROS-1* was of the Red Sea. *TIROS-1* relayed 23,000 images to Earth between April 1 and June 18, 1960. The imagery was grainy, but demonstrated it was possible to use satellites to make weather forecasts more accurate.

In 1962, the fourth craft launched in the TIROS (Television InfraRed Observation Satellite) program carried a new generation of cameras and other sensors, and the U.S. Weather Bureau began transmitting the cloud imagery to meteorological agencies worldwide.

The TIROS program continued through the 1980s. It paved the way for a suite of Earth-observing satellites that improved weather and climate research by tracking everything from rainfall rates to snow cover and sea ice, from wind speeds to the average temperature of the atmosphere.

SEE ALSO: Orbits and Ice Ages (1912), The First Computerized Forecast (1950)

**The first television picture of Earth** from space, taken on April 1, 1960, by *TIROS-1*, the first successful weather satellite.

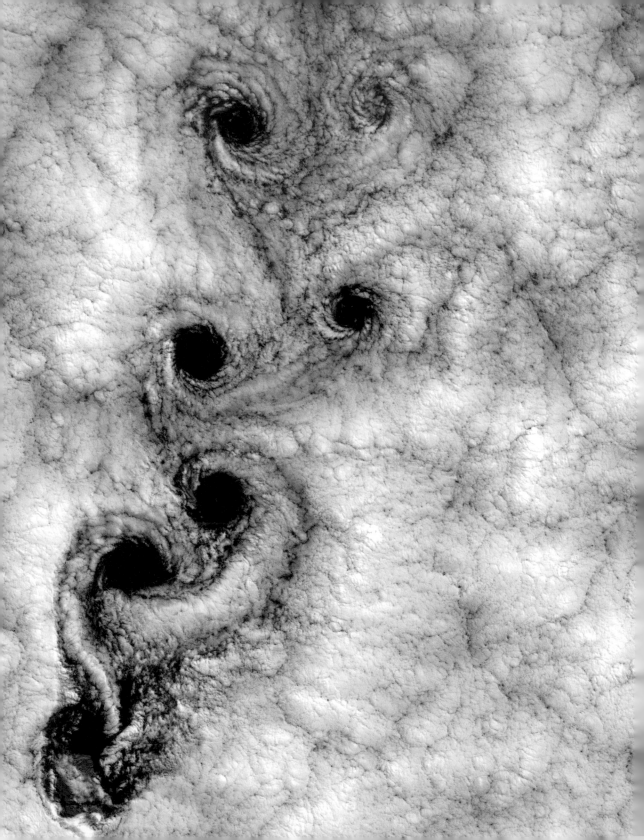

# CHAOS AND CLIMATE

B Y 1960, NUMERICAL WEATHER PREDIC-
tion—the technique laid out by Vilhelm
Bjerknes, Lewis Fry Richardson, and
others generations earlier—was advancing
almost as fast as the power of the computers.
With satellites and other sensors improving
observations and theories advancing, it was a
heady time for meteorology. Then along came
Edward N. Lorenz, and chaos theory.

Lorenz (1917–2008), a meteorologist at
Massachusetts Institute of Technology, gave
a sobering talk at a conference in Tokyo in
November 1960. He had used a simple computer
model of the atmosphere to generate weather
maps and tried to reproduce them using a standard forecasting method. The results degraded
to meaningless output beyond three days—far
faster than expected.

After the lecture the Swedish meteorologist
Bert Bolin (1925–2007) asked Lorenz if he had
tried changing the data reflecting the initial
meteorological conditions to see how much the
forecast results might vary. Lorenz said he had
changed one of the twelve variables by a fraction
of 1 percent, an amount considered insignificant.

As the historian James Rodger Fleming
recounted in his 2016 book *Inventing Atmospheric Science*:

He found that this error grew and continued to grow at a slow exponential rate until
there was no resemblance at all between
the initial and final maps. This implied
that, at least for this particular set of
equations, there is a forecasting limit. He
had identified the principle that weather
systems have a 'sensitive dependence on
initial conditions,' which is the founding
insight of chaos theory.

Lorenz intensified his research, culminating
in a landmark 1963 paper including this blunt
finding: "[P]rediction of the sufficiently distant
future is impossible by any method, unless the
present conditions are known exactly. In view of
the inevitable inaccuracy and incompleteness of
weather observations, precise very-long-range
forecasting would seem to be non-existent."

Chaos theory has since influenced fields
as varied as finance and ecology. Lorenz's
theory became known as the butterfly effect
after 1972, when a conference organizer, unbe-
knownst to him at the time, inserted this title
for his talk: "Does the Flap of a Butterfly's
Wings in Brazil Set off a Tornado in Texas?"

SEE ALSO: A "Forecast Factory" (1922), The First Computerized Forecast (1950), Climate Models Come of Age (1967)

The climate system holds both order and randomness, as can be seen in this image of a distinctive cloud
pattern called von Kármán vortex streets.

# A PRESIDENT'S CLIMATE WARNING

SPENDING ON OCEAN AND ATMOSPHERIC research was greatly boosted in the United States from the late 1950s through much of the next decade, propelled in large part by the Cold War. Early in 1965, apprised by scientists of the basics of global warming theory, President Lyndon B. Johnson (1908–73) became the first American leader to weigh in on the issue.

On February 8 of that year, Johnson delivered a written "Special Message to Congress on Conservation and Restoration of Natural Beauty," including the observation that there was "a steady increase in carbon dioxide from the burning of fossil fuels."

He warned that the environmental impacts of such pollution were no longer local and would have cumulative consequences, requiring a shift to proactive policies: "The longer we wait to act, the greater the dangers and the larger the problem. Large-scale pollution of air and waterways is no respecter of political boundaries, and its effects extend far beyond those who cause it."

While much of Johnson's focus was on noxious air and water pollution, his forward-looking recommendations presaged much of the long, difficult effort by his successors to forge national and international policy to rein in emissions of greenhouse gases.

"In addition," Johnson wrote, "the Clean Air Act should be improved to permit the Secretary of Health, Education, and Welfare to investigate potential air pollution problems before pollution happens, rather than having to wait until the damage occurs, as is now the case."

Later that year, a scientific report to the president on environmental challenges included an addendum on climate change written by a panel led by Roger Revelle and including Wallace Broecker and Charles Keeling. "The climatic changes that may be produced by the increased $CO_2$ content could be deleterious from the point of view of human beings," said the report. It even touched on the need to explore a possible remedy—"deliberately bringing about countervailing climatic changes," now known as geoengineering or climate intervention.

SEE ALSO: Global Warming Becomes News (1988), Climate Diplomacy from Rio through Paris (2015)

**President Lyndon B. Johnson** and Lady Bird Johnson walking in wildflowers near Stonewall, Texas, in 1968.

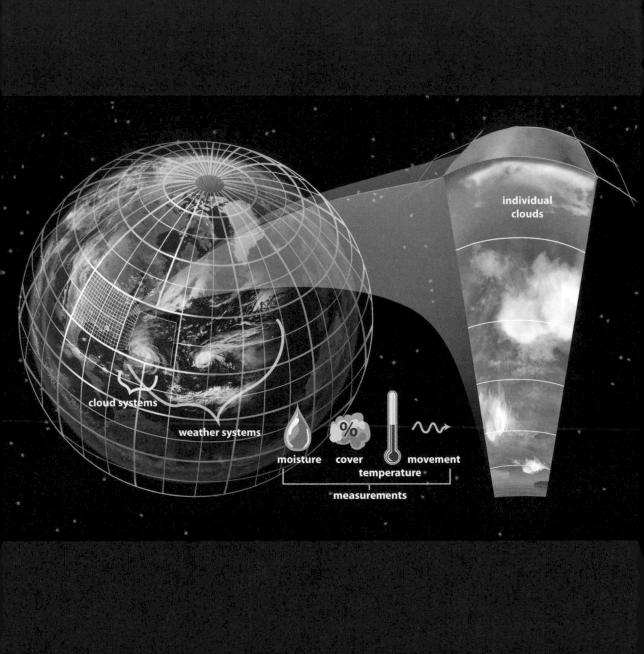

# CLIMATE MODELS COME OF AGE

ROM THE 1950S ONWARD, EVEN AS METEO-rologists refined computer-calculated forecasts of weather, efforts were undertaken to construct mathematical models that could capture the circulation of the atmosphere over longer time scales and help clarify the role of factors like the oceans, clouds, and greenhouse gases in shaping climate.

A critical first step came in 1967, with the publication of a paper describing the first attempt at using such a model to gauge how much warming could result from a substantial buildup of greenhouse gases. The one-dimensional model was exceedingly simple—representing a single column of air through the atmosphere. But it did remarkably well at estimating the sensitivity of the climate to rising concentrations of carbon dioxide.

"According to our estimate, a doubling of the $CO_2$ content in the atmosphere has the effect of raising the temperature of the atmosphere (whose relative humidity is fixed) by about 2°C [3.6°F]," wrote the authors, Syukuro Manabe (b. 1931) and Richard T. Wetherald (1936–2011) of the U.S. Geophysical Fluid Dynamics Laboratory. That figure remains close to the midpoint of dozens of estimates made since, although the range, both warmer and cooler, remains wide because of enduring sources of uncertainty such as the response of clouds in a greenhouse-heated climate.

In the paper, Manabe and Wetherald were also the first to posit (correctly) that the stratosphere would cool even as the lower atmosphere, the *troposphere*, warmed. Measurements have borne this out.

In a 1975 paper, the duo expanded their calculations into three dimensions, paving the way for general circulation models with hundreds of thousands of lines of code, run on some of the world's most powerful supercomputers, at ever-finer resolutions. In 2015, the online publication *Carbon Brief* invited authors of the most recent Intergovernmental Panel on Climate Change's report on global warming to nominate "the most influential climate change papers of all time." The 1967 paper garnered eight nominations—more than twice as many as any other study.

SEE ALSO: A "Forecast Factory" (1922), The First Computerized Forecast (1950)

**Early computer simulations of climate** represented a single column of air. Now, general circulation models are enormously detailed mathematical representations of the interactions of land, oceans, atmosphere, and ice that together shape climate.

# STORM CHASING GETS SCIENTIFIC

1973

STORM CHASING, PARTICULARLY FOCUSED on tornadoes, was not always a high-tech, high-profile effort like that popularized in films, reality television series, and live-streamed YouTube channels. One of the pioneers of storm chasing, David Hoadley (b. 1938), was an amateur weather watcher who developed his passion after a severe windstorm swept through his hometown in Bismarck, North Dakota, in 1956. Hoadley was a U.S. Army lieutenant, budget analyst for the federal government, sketch artist, and photographer. He became so focused on documenting severe weather that he arranged to take his vacations from his government job during the peak tornado season.

In the meantime, Neil Ward (1914–72), a meteorologist who worked at the National Severe Storms Laboratory in Norman, Oklahoma, from its start in 1964, brought scientific rigor to the arena, using close-up observations to develop fresh ideas about the evolution of thunderstorms and tornadoes. In 1972, the University of Oklahoma and the severe storms lab teamed up with some other researchers to start the Tornado Intercept Project, the first large-scale storm-chasing effort specifically for research.

On May 24, 1973, data collected as a major tornado developed and tore through Union City, Oklahoma, provided the foundation for studying supercell thunderstorms with persistently rotating updrafts that can generate tornadoes and microbursts. The mobile crews, along with new experimental Doppler radar at the lab, provided the first detailed data on the entire life cycle of a tornado. When the radar tapes were studied afterward, scientists could see that the rotating circulation appeared aloft before the funnel descended to the ground—showing that, at least for some tornadoes, such radar might be able to detect an early warning sign.

Doppler radar is now an unremarkable tool in severe storm tracking, and imagery captured by storm chasers is commonplace on TV and the Internet. But there's nothing ordinary about the risks. The worst day in the history of storm chasing came in 2013 when a swiftly expanding tornado changed direction and accelerated, killing three storm chasers in Oklahoma.

SEE ALSO: Franklin Chases a Whirlwind (1755), First Photographs of Tornadoes (1884), Hurricane Hunters (1943)

**A storm-chasing team** from the National Oceanic and Atmospheric Administration National Severe Storms Laboratory.

# DANGEROUS DOWNBURSTS REVEALED

TETSUYA "TED" FUJITA (1920–98), A SEVERE-storms researcher at the University of Chicago, is best known for the Fujita Scale of tornado damage that he developed in 1971 in collaboration with Allen Pearson of the National Severe Storms Forecast Center. But that tool was mainly valuable in post-storm analysis, as disaster-response teams and meteorologists tried to characterize the potency of tornadoes from the damage they left behind. Arguably, Fujita's greater contribution to public welfare was his sustained fieldwork and analysis, much of it in the face of prolonged skepticism from peers. His work revealed a deadly threat that lurked invisibly in and around some storms: concentrated downdrafts of air that could flatten forests or, more importantly, imperil aircraft as they took off or landed.

Fujita's pursuit of this deadly weather phenomenon began when he was called by an investigator of the crash of an Eastern Air Lines Boeing 727 as it landed at John F. Kennedy International Airport on June 24, 1975. There had been thunderstorms in the area, but there was no clear indication of what had knocked the plane from the sky, killing 112 passengers and injuring 12 more. Some pilots nearby had reported turbulence, while others hadn't.

The variability in the observations reminded Fujita of variations he'd seen in damage from a severe tornado outbreak the year before, including starburst-shape patterns in uprooted trees suggesting that concentrated patches of vertical winds had rushed to Earth and spread outward. For the next two years, he conducted aerial surveys, photographing and sketching such damage patterns in cornfields and forests. By 1978, he had defined and named a new phenomenon—the downburst. For downbursts covering an area smaller than 2.5 miles (4.0 kilometers) across, he proposed the name microburst.

On May 19, 1978, Fujita and researchers from the National Center for Atmospheric Research captured a vivid Doppler radar view of a microburst near Yorkville, Illinois.

**SEE ALSO:** Fastest Wind Gust (1934), Tornado Warnings Advance (1950)

**An extraordinary microburst** photographed from a helicopter over Phoenix, Arizona, on July 18, 2016.

# SEA LEVEL THREAT IN ANTARCTIC ICE

1978

A S EVIDENCE GREW IN THE LATTER HALF of the twentieth century that Earth's climate was responding to increasing concentrations of long-lived greenhouse gases emitted as a result of human activities, the prevailing assumption was that the resulting thaw of polar ice sheets and rise in seas would be substantial, but probably gradual.

Then, signs of potential ice instability and the prospect of an abrupt sea-level rise began to emerge, particularly in relation to portions of the West Antarctic Ice Sheet. This is a region on the frozen continent where sea-bottom topography is such that warm water could intrude beneath a vast volume of ice, speeding its journey to the sea.

In a paper published by *Nature* in 1978, John H. Mercer (1922–87), a glaciologist at Ohio State University, made a sobering link between human-driven climate change and the risk of abrupt loss of substantial volumes of West Antarctic ice. The paper's title minced no words: "West Antarctic ice sheet and $CO_2$ greenhouse effect: a threat of disaster."

The summary was similarly stark: "If the global consumption of fossil fuels continues to grow at its present rate, atmospheric $CO_2$ content will double in about 50 years. Climatic models suggest that the resultant greenhouse-warming effect will be greatly magnified in high latitudes. The computed temperature rise . . . could start rapid deglaciation of West Antarctica, leading to a 5-meter [16-foot] rise in sea level."

Mercer was long seen as an outlier, but his views have gained support of late. Two independent studies in 2014 posited that the West Antarctic "collapse" was now inevitable—albeit on a time scale still measured in centuries rather than decades or years. And of course Antarctica is only one source of volumes of water sufficient to threaten today's coastal cities. Greenland has a far smaller ice sheet, but its two-mile-high (three-kilometer-high) ice cap holds water equal to that of the Gulf of Mexico.

**SEE ALSO:** The Coldest Place on Earth (1983), Climate Clues in Ice and Mud (1993), Arctic Sea Ice Retreat (2016)

**A portion of the Thwaites Glacier**, pictured in 2012. This is the leading edge of ice flowing into the sea from part of the West Antarctic Ice Sheet.

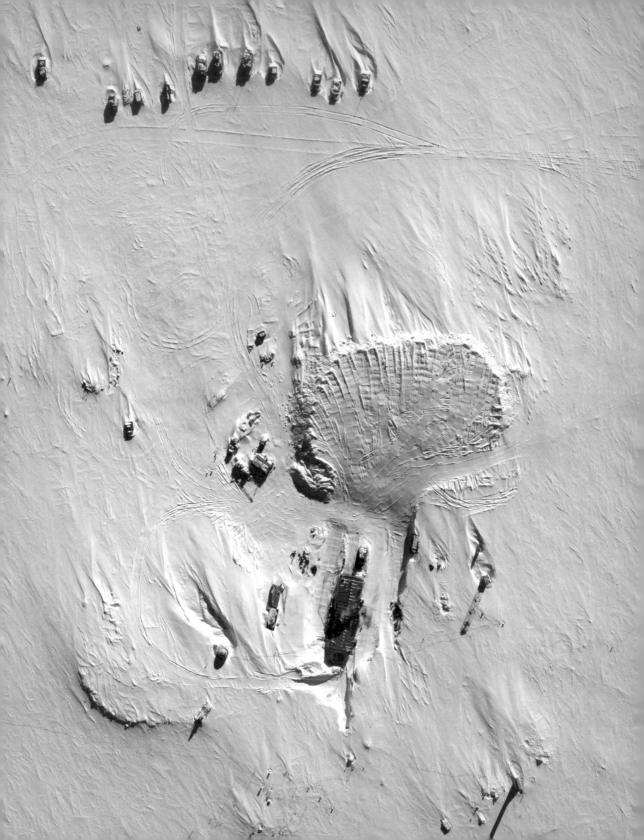

# THE COLDEST PLACE ON EARTH

ON JULY 21, 1983, THE THERMOMETER at Russia's Vostok research station in eastern Antarctica recorded a temperature of –128.6 degrees Fahrenheit (–89.2 degrees Celsius). This is the coldest temperature ever measured with a thermometer, making the *Guinness Book of World Records* and set as the record by the World Meteorological Organization. The location, at an elevation of 11,444 feet (3,488 meters) on the vast interior ice sheet, is informally called the Pole of Cold.

There may be colder spots. On August 10, 2010, researchers from the National Snow and Ice Data Center in Boulder, Colorado, announced that they had recorded the unfathomably frigid temperature of –135.8 degrees Fahrenheit (–93.2 degrees Celsius) in Antarctica's eastern highlands. The researchers from the National Snow and Ice Data Center studied over thirty years of global surface temperature maps using data from remote sensing satellites. They unexpectedly discovered record lows on a high ridge between Dome Argus and Dome Fuji. (They had assumed the coldest air, being the densest, would be in the lowest spots.)

Through Russian media, the head of the Russian Antarctic Expedition logistics center protested, saying it was "incorrect and unrealistic" to declare this record based on satellite data. The Russians had protocol on their side, in any case. To break the record, the temperature had to be recorded by thermometers, according to international meteorological protocol.

As with the hottest spots on the planet, it's worth noting the record for places people live. According to NASA, the coldest permanently inhabited place on Earth is northeastern Siberia, where temperatures dropped to 90 degrees below zero Fahrenheit (–67.8°C) in the towns of Verkhoyansk (in 1892) and Oimekon (in 1933).

**SEE ALSO:** Settling a Hot Debate (2012), The Polar Vortex (2014)

**In 1983, the coldest temperature** measured with a thermometer was recorded at Russia's Vostok Station, about 800 miles (1,287 kilometers) from the South Pole.

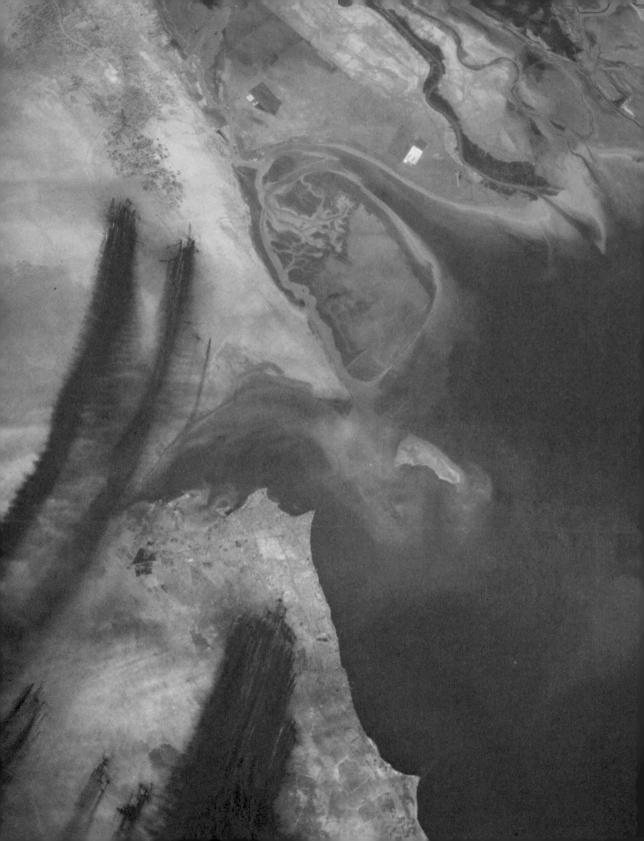

# NUCLEAR WINTER

A PERFECT STORM OF COLD WAR, ENVIronmental concerns, and evolving climate science erupted in the early 1980s, producing portentous warnings of a new kind of environmental threat—a potential "nuclear winter" triggered by great conflagrations following a nuclear exchange.

Vast dense smoke clouds rising from cities burned in a nuclear war could chill the planet, triggering famine and worse. The concept evolved through a series of early papers, starting with "The Atmosphere After a Nuclear War: Twilight at Noon," a 1982 analysis by two atmospheric scientists, Paul J. Crutzen (b. 1933) and John W. Birks (b. 1946). Crutzen had risen to prominence for research in the early 1970s identifying chemical reactions that could weaken the planet's protective ozone layer—work that would earn him a share of the Nobel Prize in chemistry in 1995.

But the nuclear winter hypothesis gained greatest visibility through the involvement of Carl Sagan (1934–1996), who was one of five authors of a December 23, 1983, *Science* paper, "Nuclear Winter: Global Consequences of Multiple Nuclear Explosions." Sagan alerted millions of people to the threat in a *Parade* magazine special report and in television appearances. He and a Soviet counterpart, Vladimir Alexandrov (1938–85), made a Vatican visit and other appearances, seeking a ban on nuclear weapons.

Under deeper scientific scrutiny, the initial apocalyptic scenarios grew more nuanced, with another prominent climate scientist, Stephen H. Schneider (1945–2010), seeing more of a "nuclear autumn." With the collapse of the Soviet Union, the threat of nuclear war ebbed.

But more recent climate simulations by Alan Robock (b. 1949) and Owen Brian Toon (b. 1947), one of the authors on the 1983 paper with Carl Sagan, point to a decade of devastating climatic disruption from even a limited nuclear exchange. The reason? Smoke from nuclear conflagrations could rise to an altitude of 25 miles (40 kilometers), far too high to be washed out promptly by precipitation.

SEE ALSO: Russia's "General Winter" (1941), Climate by Design? (2006)

**In the early 1980s**, scientists calculated that hundreds of fires ignited in a nuclear war could loft dark, sun-blocking smoke clouds and chill the planet in a "nuclear winter." In 1991, as Saddam Hussein ignited Kuwaiti oil wells, scientists measured local cooling from the spreading black clouds, but the smoke did not rise high enough to have a wider effect.

# FORECASTING EL NIÑO

1986

HERE IS NO PERIODIC WEATHER EVENT ON the planet with as widespread an impact, for better or worse, as the El Niño-Southern Oscillation. This phenomenon is an irregular shift in Pacific Ocean temperatures that can stifle Atlantic hurricanes even as it spawns Indonesian wildfires; shift patterns of drought and heavy rain; bleach coral reefs; and much more.

The Spanish part of the name came from Peruvian geographers who, in the late 1800s, described a warm "El Niño counter-current" that fishermen along the coast said occasionally ruined anchovy catches, often around Christmastime (thus, the reference to "the boy child"). It took a string of scientists decades to discern the forces at play, realize the cycle's global reach, and then craft models leading to useful forecasts. The first deciphered component was a periodic seesawing shift in atmospheric pressures over the Pacific and Indian Oceans. The pattern emerged in 1923 in statistical studies of global weather data by Gilbert Walker (1868–1958), a mathematician based in India who was trying to find atmospheric patterns that could explain the occasional years in which South Asia's life-giving monsoon rains fail to occur.

In 1969, Jacob Bjerknes (1897–1975) of the University of California, Los Angeles, connected the atmospheric cycle to warm and cool phases of the tropical Pacific. Two more scientists, George Philander (b. 1942) of Princeton University and Mark Cane (b. 1944) of Columbia University, worked out how tropical winds and currents could sometimes create the self-reinforcing warming pattern, and an opposite cool counterpart, which Philander dubbed La Niña, "the girl child," in 1985.

That year, Cane and a student, Stephen Zebiak, developed a forecast model coupling oceanic and atmospheric data. In a paper published in June 1986, they successfully predicted the emergence of an El Niño episode. Many other models have been developed since. The cycle still sometimes confounds the experts, but communities now generally have much more time to plan for this disruptive pattern.

**SEE ALSO:** Tracking the Oceans' Climate Role (2007), Reefs Feel the Heat (2017)

**An extremely strong El Niño warming** in the tropical Pacific in 1997 and 1998 had widespread impacts, including flooding along California's Russian River in March 1998.

# GLOBAL WARMING BECOMES NEWS

FROM THE LATE 1800S ONWARD, DRIVEN by the findings of Sweden's Svante Arrhenius and successors, there was a sprinkling of news coverage of the basic hypothesis that carbon dioxide emitted by fuel burning could warm the planet's climate. A 1912 item, originally printed in *Popular Mechanics* and reproduced in newspapers as far afield as Australia, concisely summarized the basics:

> The furnaces of the world are now burning about 2,000,000,000 tons [1,814,000,000 tonnes] of coal a year. When this is burned, uniting with oxygen, it adds about 7,000,000,000 tons [6,350,000,000 tonnes] of carbon dioxide to the atmosphere yearly. This tends to make the air a more effective blanket for the earth and to raise its temperature. The effect may be considerable in a few centuries.

Of course, the pace of emissions far exceeded early projections as demand for coal surged and was matched by the rise in the use of oil and natural gas, driven by expanding populations, transportation, industrial production, and electricity use. Science continued to refine the basic picture of a rising human influence on climate, particularly from the late 1950s on.

In 1988, building on global concerns about deforestation, acid rain, and damage to the ozone layer from certain synthetic chemicals, global warming jumped from an esoteric news item to front pages. On June 23, a NASA climate scientist, James Hansen (b. 1941), told a U.S. Senate committee that human-produced greenhouse gases were measurably heating the climate.

Hansen had stepped out ahead of most peers, but there were plenty of cues from nature, as well, including a record North American heat wave and a wildfire in Yellowstone National Park.

That summer, scientists and diplomats gathered in Canada for the "Toronto Conference on the Changing Atmosphere," and recommended global reductions in greenhouse-gas emissions. The Intergovernmental Panel on Climate Change was also formed that year under the auspices of the United Nations to advise the world's nations on climate risks and responses.

SEE ALSO: Coal, $CO_2$, and Climate (1896), A President's Climate Warning (1965), Climate Diplomacy from Rio through Paris (2015)

**Global warming first became a big news story** in 1988, with record heat and fires from the Amazon rain forest to Yellowstone National Park (seen here) making headlines.

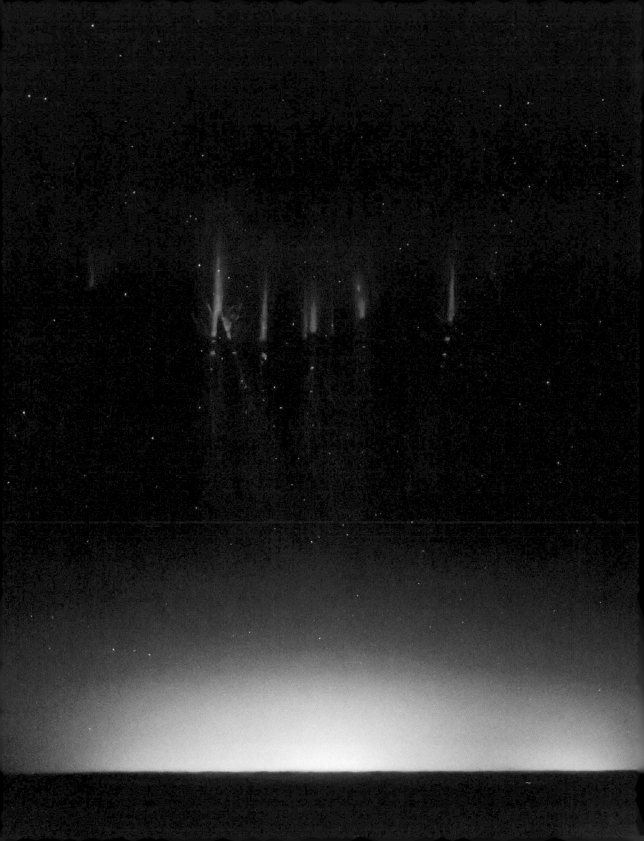

# PROOF OF ELECTRICAL "SPRITES"

IN 1973, U.S. AIR FORCE PILOT RONALD WILLIAMS was flying over a typhoon in the South China Sea. As he flew close to a thunderstorm near the center of the giant storm, he saw something like a bolt of lightning go straight up from the top of the clouds. When Williams reported his sighting, he was told that lightning doesn't go up—it has to discharge on something. Through much of the twentieth century, high-flying military and civilian pilots had reported phenomena of this sort.

But it wasn't until 1989 that a chance sighting provided firm visual evidence. With a small team, John Winckler (1916–2001), a physicist from the University of Minnesota, was testing a low-light video camera and accidentally captured a black-and-white image of upward-streaking electrical discharges. The name "sprite" was proposed in the mid-1990s by Davis Sentman (d. 2011), leader of a University of Alaska Geophysical Institute team videotaping thunderstorms in the Midwest from two high-flying NASA aircraft.

Lightning is a hot discharge of energy with a negative electric charge. Sprites have a positive charge and are more akin to the cold glow in a fluorescent bulb. The electric charge from sprites is ten times more powerful than the average lightning bolt. This can cause the energy from a sprite to reverberate measurably around the world.

Scientists have described three kinds of sprites: jellyfish sprites, which can spread up to 30 by 30 miles across; column sprites, with large-scale electrical discharges; and carrot sprites, vertical red columns with electrical tendrils.

The most dramatic and detailed photographs of sprites have been taken by astronauts aboard the International Space Station showing that some can reach as high as 60 miles (97 kilometers) above the Earth's surface—well into the ionosphere.

SEE ALSO: Deciphering the Rainbow (1637), Extreme Lightning (2016)

**An elusive electrical discharge called a sprite**, photographed in 2013 from a high-flying research jet by researchers from the University of Alaska Fairbanks.

# CLIMATE CLUES IN ICE AND MUD

CRITICAL LINES OF EVIDENCE CONFIRMing past cycles of ice age cold and intervening warmth emerged in the twentieth century from cylinders of seabed mud and glacial ice. One key hint in layered seabed sediment was variations in trace oxygen isotopes reflecting different temperatures. In 1976, a landmark analysis of 450,000 years of seabed evidence by three scientists, titled "Variations in the Earth's Orbit: Pacemaker of the Ice Ages," supported the findings of Milutin Milanković fifty years earlier.

Ice cores from Greenland and Antarctica proved a particularly invaluable trove, with bubbles of ancient air showing past concentrations of the greenhouse gases carbon dioxide and methane, along with ash from volcanoes, soot from forest fires, and other indicators of past environmental changes.

A particularly important discovery came in 1993, when scientists analyzing the ice-core records published evidence that extraordinarily abrupt climate jogs had occurred—likely driven by changes in ocean circulation. Richard B. Alley (b. 1957), a leader of the Greenland research effort, describes the insight this way:

> The warming and cooling sometimes crossed thresholds that shifted how the ocean circulates, and whether the wintertime North Atlantic is open water or sea ice far below freezing. The climate sometimes jumped between those states in years rather than decades, with impacts that spread around the globe. We remain fairly confident that warming won't melt Greenland's ice fast enough to trigger such a jump in our near future. But, the climate history is clear—raising $CO_2$ will impact climate and living things in major ways, with some chance of even larger, more disruptive events on the way to a warmer future.

Many experts in risk management point to evidence for abrupt, and poorly understood, changes in climate as justifying efforts to curb heat-trapping emissions.

SEE ALSO: Orbits and Ice Ages (1912), Climate Models Come of Age (1967)

**Hundreds of cylinders of ancient ice** holding climate clues are stored at the National Ice Core Laboratory in Lakewood, Colorado.

# THE HUMAN FACTOR IN WEATHER DISASTERS

FOLLOWING THE RECORD-BREAKING ATLANtic hurricane season of 2005, in which Katrina's devastating blow to New Orleans was a focal point, debate intensified among climate scientists over how much and how soon global warming might influence such storms. Hurricanes also became an icon in environmental politics.

Lost in the disputes was an unnerving reality: rapid and poorly planned growth in coastal communities was increasing the storm threat with or without warming. On July 25, 2006, a group of leading hurricane researchers with varied views on warming's role in fueling storms issued a joint statement for media and the public. The statement—signed by Kerry Emanuel, Richard Anthes, Judith Curry, James Elsner, Greg Holland, Phil Klotzbach, Tom Knutson, Chris Landsea, Max Mayfield, and Peter Webster—centered on this point:

> While the debate on this issue is of considerable scientific and societal interest and concern, it should in no event detract from the main hurricane problem facing the United States: the ever-growing concentration of population and wealth in vulnerable coastal regions. These demographic trends are setting us up for rapidly increasing human and economic losses from hurricane disasters, especially in this era of heightened activity. Scores of scientists and engineers had warned of the threat to New Orleans long before climate change was seriously considered, and a Katrina-like storm or worse was (and is) inevitable even in a stable climate. . . . We call upon leaders of government and industry to undertake a comprehensive evaluation of building practices, and insurance, land use, and disaster relief policies that currently serve to promote an ever-increasing vulnerability to hurricanes.

Notably, the basic message could apply just as well in communities prone to wildfire, tornadoes, and inland flooding. The inundation of Houston and neighboring regions after record rainfall from Hurricane Harvery in August 2017 revived such calls.

SEE ALSO: A Mighty Storm (1900), "China's Sorrow" (1931), Hurricane Hunters (1943)

In 2006, ten hurricane researchers warned that rapid coastal development was greatly increasing risk from such storms. This view from a U.S. National Guard helicopter shows the consequences after Hurricane Sandy swamped New Jersey communities in 2012.

# CLIMATE BY DESIGN?

FTER GENERATIONS OF EFFORTS TO modify weather, most of which were more theatrical than effective, some scientists from the mid-twentieth century onward began considering ways to influence the climate system itself—in an emerging field most widely known today as geoengineering. A 2015 National Academy of Sciences report chose the term *climate intervention* to describe such efforts.

As various ways to counteract warming were proposed, debate sprang up as well. As the historian James Rodger Fleming chronicled, Harry Wexler (1911–62), the director of research at the U.S. Weather Bureau, warned as early as 1962: "Most such schemes that have been advanced would require colossal engineering feats and contain the inherent risk of irremediable harm to our planet or side effects counterbalancing the possible short-term benefits."

The focus on the field sharpened in 2006 when the journal *Climatic Change* published an essay in support of such research by Paul J. Crutzen, who'd shared a 1995 Nobel Prize with Mario J. Molina and F. Sherwood Rowland for identifying chemical threats to Earth's protective ozone layer. Lamenting that calls for reductions in climate-warming emissions appeared to be little more than a "pious wish," he proposed ramping up research on ways to cool things.

There have been two main paths of inquiry aimed at averting undesired warming. One, called solar radiation management, centers on ways to add bright particles to the atmosphere, changing the planet's albedo (or, reflectivity) by blocking some incoming sunlight in the same way high-altitude plumes from big volcanoes and some kinds of air pollutants have done.

Other efforts are focused on capturing carbon dioxide from the atmosphere or smokestack emissions and storing it underground, or breaking down $CO_2$ molecules and locking the carbon in stable materials. In another approach, iron dust has been used to fertilize ocean plankton in ways that might deposit the carbon on the seabed. There are a host of issues, ranging from technical to ethical to diplomatic. Perhaps the biggest issue is simply the scale of emissions—tens of billions of tons of liberated $CO_2$ a year.

SEE ALSO: Rainmakers (1946), A President's Climate Warning (1965), An End to Ice Ages? (102,018 CE)

---

**Some scientists have proposed testing ways** to create a sun-blocking haze to counter global warming from the buildup of greenhouse gases in the atmosphere. These are contrails over Germany.

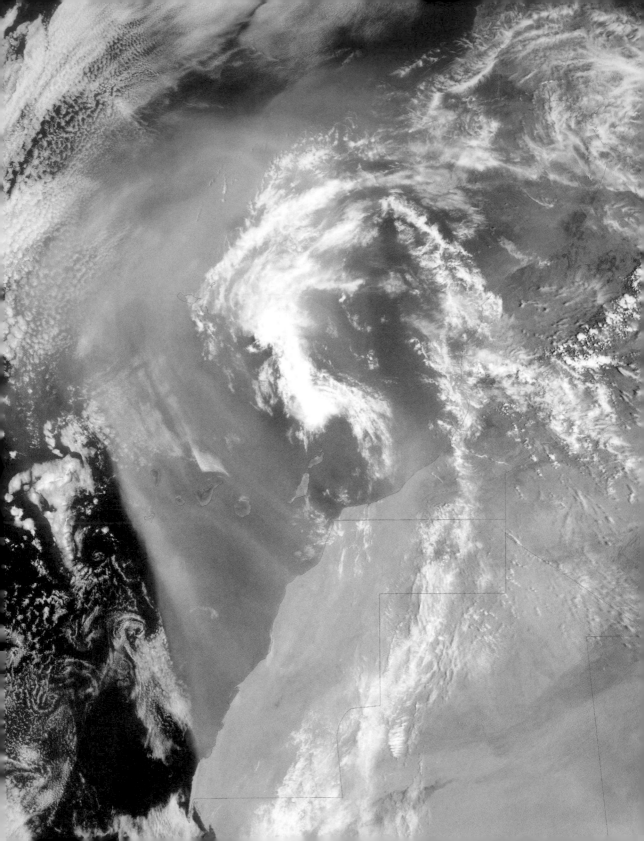

# LONG-DISTANCE DUST

THERE IS PERHAPS NO SINGLE EXAMPLE of the powerful interconnections between Earth's dynamic atmosphere, landscapes, and ecosystems than the ocean-spanning influence of sandstorms in the Sahara Desert—on everything from hurricanes to Brazilian forests to Bahamian beaches.

Various studies in the early 2000s revealed that a dry, dusty Sahara Air Layer in the atmosphere had the capacity to stifle hurricanes, adding a fresh variable to the other factors that determine the tropical storm threat in any particular season. (As another example, Pacific Ocean El Niño hot spells shift atmospheric circulation in ways that also blunt Atlantic hurricanes.)

Around the same time, research meshing satellite monitoring and rain forest ecology was revealing that nutrient-rich dust from the Sahara played a role in fertilizing the rain forests of the Amazon. A particularly notable 2006 study led by scientists at Israel's Weizmann Institute of Science estimated that 40 million tons (36.3 million tonnes) of North African dust fell on the Amazon each year, with more than half of the material coming from a single windswept spot in northern Chad—the Bodélé Depression, an ancient dry lakebed flanked by basalt ridges that create a natural wind tunnel.

But dust from North Africa had a very different impact farther north. In 2014, University of Miami researchers studying iron and manganese levels in 270 samples pulled from the shallow white limestone banks in the Bahamas concluded that Saharan dust fertilizing blooms of photosynthesizing cyanobacteria was instrumental in building the foundation on which those islands sit today.

In 2016, other researchers probing cores from the Bahamas found that this fertilizing dust flow varied enormously in the past. In the latter stages of the last ice age, 23,000 years ago, twice today's volume of dust fell around the Bahamas. But from 11,000 to 5,000 years ago, the plumes carried half as much dust as today.

SEE ALSO: North Africa Dries and the Pharaohs Rise (5,300 BCE), The Dust Bowl (1935), Forecasting El Niño (1986)

**A 2012 image showing dust from the Sahara** spreading west over the Atlantic Ocean. Such plumes affect regions as distant as the Amazon rain forest.

# TRACKING THE OCEANS' CLIMATE ROLE

THE OCEANS, COVERING TWO-THIRDS OF the planet's surface, play a profound role in shaping the climate at the global and regional level, storing and moving massive amounts of solar energy, and serving as the source of nearly 90 percent of the evaporation that determines humidity and ultimately shapes cloud and precipitation patterns.

Yet despite their importance, the oceans were a data desert until recently. Over the twentieth century, the global spread and integration of thousands of land-based weather stations, and then the launch of strings of satellites, greatly advanced both weather and climate understanding. Still, critical ocean trends had to be estimated by sifting through reams of inconsistent observations from ships, moored buoys, submarines, and the like.

That all changed in 1999, when scientists convinced dozens of nations to contribute to Argo, a shared international system for tracking undersea ocean temperatures, salinity, and currents, with such data critical to improving everything from global warming models to El Niño and hurricane forecasts.

In November 2007, Argo became fully operational. It's a network of nearly 4,000 widely dispersed autonomous buoys that drift at a depth of over half a mile (805 meters), and then—every ten days—descend to a depth of 1.2 miles (1.9 kilometers) and rise back to the surface, collecting a profile of temperature and salinity measurements that is then transmitted via satellite.

Each year, hundreds of research papers are published drawing on Argo data. Among other insights, the system in 2017 helped clarify that the warming of the climate did not unexpectedly pause in the early 2000s, as some lines of evidence hinted. While politicized debates over global warming policy centered for a while on the idea of a pause, or hiatus, this variation is now widely seen as one of many instances in which atmospheric temperatures fluctuated on the long-term journey to a warmer planet.

You can learn more at the Argo website, argo.ucsd.edu, maintained by the Scripps Institution of Oceanography, where much of the technology behind the system was developed.

SEE ALSO: First Weather Balloon Flight (1783), Meteorology Gets Useful (1870), Watching Weather from Orbit (1960)

**An Arvor Iridium deep-diving data-gathering buoy**, one of several thousand such devices distributed around the world's oceans to track conditions for the Argo System.

# SCIENCE PROBES THE POLITICAL CLIMATE

THROUGH THE EARLY YEARS OF THE TWENTY-first century, efforts intensified around the world to restrict greenhouse gases. In the United States, this push led to intensifying polarization around the issue, although surveys showed most of the public was largely disengaged at the same time. Psychologists and social scientists began to examine what was shaping this paralyzing mix of detachment and division. The incremental nature of climate change, unfolding in ways still mostly obscured by natural variability in the climate system, was a bad fit for deep-rooted human traits, including what behavioral scientists call our "finite pool of worry"—i.e., the climate can't compete with bills, kids, health concerns, and so on.

In 2012, an eye-opening study in the journal *Nature Climate Change* led by Dan Kahan, a law and psychology professor at Yale University, described a telltale pattern he calls cultural cognition, in which it is more rational for someone to maintain cultural ties than to accept facts that clash with that identity. This and subsequent studies used a mix of survey questions and experiments to reveal people's basic scientific knowledge; worldview—in essence, testing if you are a group hugger or a loner, among other traits; and reactions to information. The studies showed that the greatest understanding of the basics of climate science was found in people at both ends of the spectrum of concern about climate change—those most alarmed and most dismissive.

In a commentary published in *Nature* later that year, Kahan gave this example: "People with different values draw different inferences from the same evidence. Present them with a Ph.D. scientist who is a member of the U.S. National Academy of Sciences, for example, and they will disagree on whether he really is an 'expert,' depending on whether his view matches the dominant view of their cultural group."

According to Kahan, these sobering findings are not a dead end. He has worked with politically divided communities in hot, flood-prone southeastern Florida that have found ways to pursue policies that boost resilience to weather extremes or that develop nonpolluting energy sources that can suit more than one ideology.

SEE ALSO: Global Warming Becomes News (1988), Climate Diplomacy from Rio through Paris (2015)

**Protest signs at the 2009 climate treaty talks in Copenhagen.** Behavioral scientists have found that people with strong, but divergent, political stances can have a good understanding of climate science but completely different views of climate change risks.

# SETTLING A HOT DEBATE

ARLY IN 2010, CHRISTOPHER BURT, author of *Extreme Weather: A Guide and Record Book*, received a provocative email questioning his listing for the hottest measured temperature—136.4 degrees Fahrenheit (58 degrees Celsius), recorded on September 13, 1922, at an Italian fort outside El Azizia, a trading post in what is now Libya.

The El Azizia listing had seemed safely ahead of the next hottest figure—134 degrees Fahrenheit (56.7 degrees Celsius), measured on July 10, 1913, at Greenland Ranch in Death Valley, California. But in the email, Maximiliano Herrera, an authority on temperature measurements, called it a "garbage record."

A wider conversation ensued. Burt reached out to Khalid Ibrahim El Fadli, the director of the climate department at the Libyan National Meteorological Center. "I asked him," Burt recalled in an interview for this book, "'This is your country, your data, do you believe it?'" The answer was a flat no.

El Fadli began to uncover original documents. That fall, Burt wrote a post for the Weather Underground website challenging the record. The World Meteorological Organization convened a panel including El Fadli, Burt, and other experts. A host of problems surfaced, including the use of a finicky thermometer and the discovery, in a logbook found by El Fadli, that there had been a personnel change at the suspect site just before readings there sharply diverged upward compared to temperatures recorded at other sites in the region.

In September 2012, the World Meteorological Organization shifted the record for hottest measured temperature to the 1913 Death Valley reading. But the story continues. Burt and others have said that reading is almost surely invalid, as well. What is clear, according to Burt, is the value of consistent measurement.

What is also clear, from recent research in North Africa and the Middle East, is that what counts as record heat these days will become far more normal later in this century should greenhouse-driven warming continue.

**SEE ALSO:** A Dry Discovery (1903), The Coldest Place on Earth (1983)

**A photograph of Dante's View in Death Valley**, California. In 2012, the world record for hottest temperature was shifted by the WMO from a measurement in North Africa in 1922 to one in Death Valley in 1913.

# THE POLAR VORTEX

O N TUESDAY, JANUARY 7, 2014, FIFTY weather stations across the United States measured record cold temperatures for that date. A large part of the country, from Montana to New York, and south to Oklahoma and Alabama, recorded sub-zero temperatures. Life-threatening wind chills across the Midwest made it feel like 40 degrees below zero. Newscasters excitedly announced a disruption of the polar vortex, which is a band of counterclockwise high-altitude winds around the North Pole, as the cause, pointing to weather charts showing big wiggles in the jet stream and bulging blue zones of sub-zero air.

The phrase has surfaced periodically since—almost any time similar outbreaks of Arctic air spread south across Europe or North America, most accompanied by surges of warm air into the Arctic. But the polar vortex is hardly new, as Jeff Masters, one of the meteorologists who founded the popular Weather Underground website, explained that winter in an attempt to quell a viral burst of scary headlines.

"This meteorological phenomenon has likely been present for as long as there has been weather on Earth," he wrote. He also noted that the phrase has been in scientific papers at least since a 1939 study by Carl-Gustaf Arvid Rossby (1898–1957), a pioneering Swedish-born meteorologist who transformed our understanding of large-scale circulation patterns in the atmosphere.

Some recent studies have suggested that such cold outbreaks and even patterns in extreme winter storms could be related to human-driven climate change, with reductions in Arctic Ocean sea ice in recent years possibly playing a role.

But most scientists probing links between weather patterns in the Northern Hemisphere and climate change have said it is premature to draw clear conclusions given the extraordinary mix of causes and effects shaping weather in and around the Arctic and the limited span of decades with precise measurements using satellites and other sensors.

SEE ALSO: The Great White Hurricane (1888), The Great Blue Norther (1911), The Coldest Place on Earth (1983)

**A potent cold front advances** over the northeastern United States on February 14, 2016, in this photograph taken from the International Space Station.

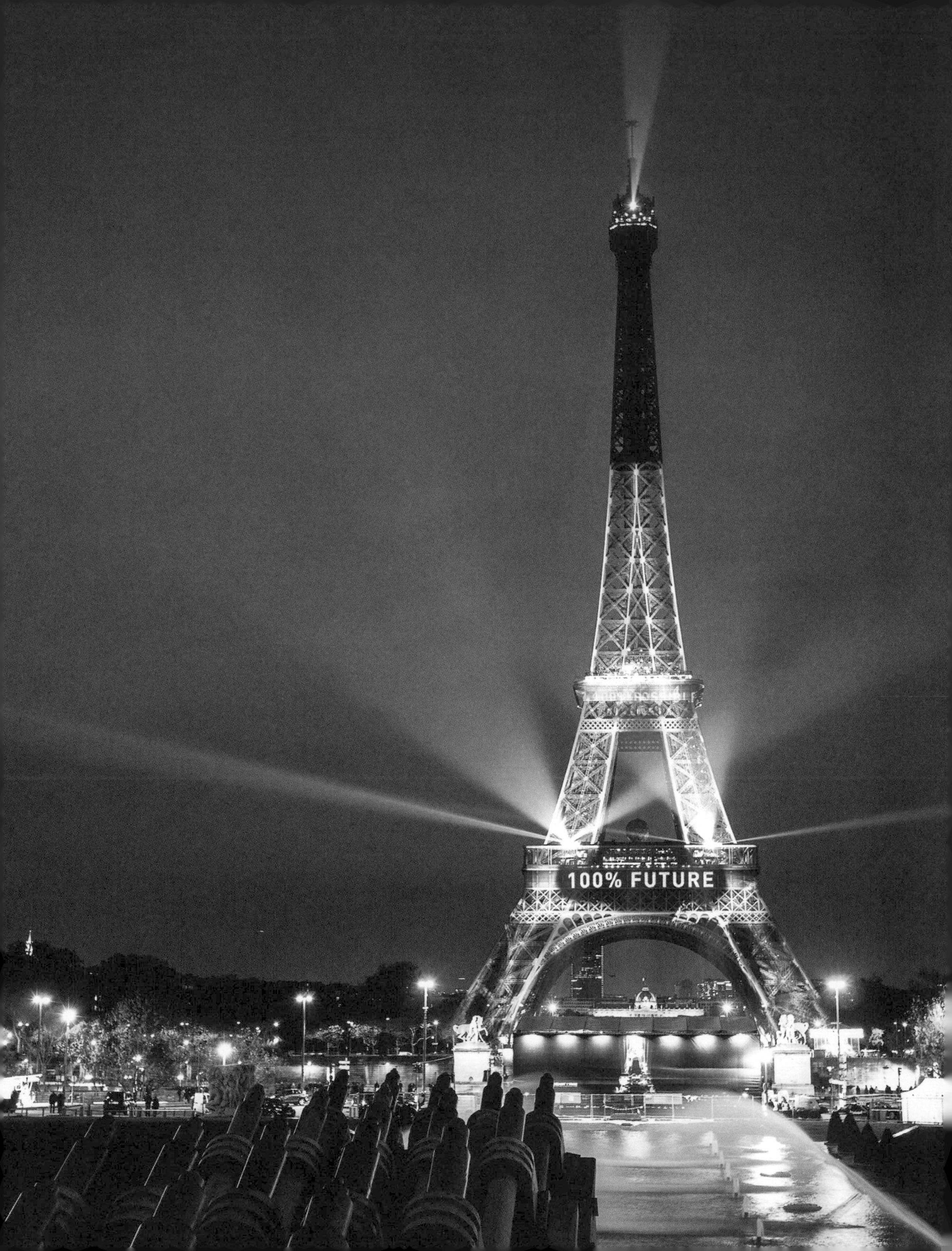
100% FUTURE

# CLIMATE DIPLOMACY FROM RIO THROUGH PARIS

WITH A GAVEL BANG IN PARIS ON December 12, 2015, officials from 195 nations adopted the first international climate agreement in which nearly all countries—rich and powerful, poor and obscure—pledged to take steps to avoid dangerous human-driven warming by reducing emissions of greenhouse gases.

Even with the cheering and headlines, the impact of the resulting Paris Agreement was always going to be limited because it included only voluntary steps on emissions and pledges by rich nations to help finance resilient development and clean energy projects in poor ones. And momentum appeared to weaken in June 2017, when U.S. President Donald J. Trump pledged to withdraw the United States from the agreement. But that process itself would take years and Trump appeared to hedge, hinting that some renegotiation might be possible.

In the end, focusing on the Paris pact—or on any particular country or politician's decisions—as make-or-break moments in slowing global warming misses a couple of important realities.

The Paris Agreement is just one step in a continuing journey that began when 196 countries adopted the original (and also nonbinding) global warming treaty, the United Nations Framework Convention on Climate Change, at the Earth Summit in Rio de Janeiro, Brazil, in 1992. That pact was signed by President George H. W. Bush, a Republican, and ratified later that year by the U.S. Senate. It was supported by three successive presidents, including Bill Clinton, George W. Bush, and Barack Obama.

When it comes to energy and economies, diplomacy and politics mostly reflect more than determine what nations and people choose to do. Coal and oil, the main sources of carbon dioxide emissions, will still be in the global energy mix for decades to come. But technological advances that have opened vast new reserves of cleaner natural gas, cut costs of renewable energy, and pointed to new nuclear power plant designs could, with continued investment and effort, signal a shift to a more sustainable human relationship with the Earth's climate.

SEE ALSO: A President's Climate Warning (1965), Global Warming Becomes News (1988)

**The 2015 Paris Agreement** was the first climate accord in which developed and developing countries all pledged to pursue cuts in greenhouse gas emissions.

# ARCTIC SEA ICE RETREAT

THE CLIMATE AND ICE CONDITIONS AROUND the North Pole have gone through profound changes through the planet's history. A 2004 seabed drilling project there discovered that some 55 million years ago, when the global climate was at a hot peak driven by a dramatic buildup of greenhouse gases, the sea surface was a near-tropical 74 degrees Fahrenheit (23.3 degrees Celsius). Around 49 million years ago, mats of duckweed coated Arctic waters. But the same research found that a sheath of sea ice appeared to have cloaked the region consistently from 15 million years ago until recently.

Of course, things appear to be changing dramatically once again. The Arctic Ocean still freezes up each dark, frigid boreal winter, but sea ice has retreated so much in recent summers that shipping is becoming unremarkable in areas that were impassable 200 years ago.

Since 1979, satellites have measured a trend toward more open water, leading to predictions of a largely ice-free Arctic Ocean in late summers sometime later this century—although with lots of variations possible along the way.

In 2016, a pioneering study, sifting through a host of old and scattered records including whaling ship logs and old Soviet ice surveys, compiled a sea ice chronology going back to 1850. Florence Fetterer, one of the authors, summarized the findings succinctly:

> First, there is no point in the past 150 years where sea ice extent is as small as it has been in recent years. Second, the rate of sea ice retreat in recent years is also unprecedented in the historical record. And, third, the natural fluctuations in sea ice over multiple decades are generally smaller than the year-to-year variability.

The study added to a growing body of science pointing to Arctic warming from the buildup of human-generated greenhouse gases as the dominant cause.

SEE ALSO: Cold Dooms an Arctic Explorer (1845), Sea Level Threat in Antarctic Ice (1978)

**Scientists in 2016 published research**, conducted by sifting through a host of old records including logs of whaling ships, which confirmed the unusual nature of recent retreats in Arctic sea ice in summer. This 1871 print depicts whaling in the Atlantic Ocean and Bering Strait.

# EXTREME LIGHTNING

IGHTNING STRIKES SOMEWHERE ON EARTH every second, as satellites now routinely record. But only one place is considered the lightning capital of the world. Until recently, the Congo Basin in Africa held that title. But a 2016 analysis by NASA scientists, using sixteen years of detailed lightning data collected by satellite-based instruments, produced a new record holder: Lake Maracaibo in northwest Venezuela. Maracaibo, the largest lake in South America, is fringed on the south by a horseshoe array of mountains that trap warm, humid trade winds flowing from the Caribbean to the north. These winds meet cool air coming down from the Andes, forming massive thunderheads. This persistent condition results in an average of 297 nocturnal thunderstorms a year, peaking in September. The conditions at Lake Maracaibo are sufficiently dramatic that a nearby camp specializes in lightning tourism.

In 2010, possibly as a result of a severe drought linked to an El Niño episode, locals were startled when the lightning displays stopped for several months—marking possibly the longest lightning hiatus there in a century. But the flashes resumed.

At the height of the wet season, there are an average of twenty-eight lightning flashes per minute. The satellite lightning sensors, developed and managed at the space agency's Marshall Space Flight Center in Huntsville, Alabama, provide a fine-grained look at the density and frequency of lightning worldwide. The sensors are sensitive enough to spot lightning flashes even in the daytime.

The Marshall Center was also involved in determining the record for the length of a lightning flash—a bolt spanning 199.5 miles (321 kilometers) of sky over Oklahoma in 2007. The World Meteorological Organization announced that record in 2016, along with the record for the longest duration for a single lightning flash—7.74 seconds—in southern France in 2012.

SEE ALSO: Benjamin Franklin's Lightning Rod (1752), Dangerous Downbursts Revealed (1975), Proof of Electrical "Sprites" (1989)

**In 2016, a NASA analysis of satellite imagery** determined that Venezuela's Lake Maracaibo, due to a mix of geographic and meteorological conditions, is the lightning capital of the world.

# REEFS FEEL THE HEAT

FOR MORE THAN 200 MILLION YEARS, THE coral animals that build most of the world's tropical reefs have had a mutually beneficial relationship with certain algae. Through photosynthesis, the algae, sheltered in coral polyp tissues, provide nutrients to the corals, which emit certain waste products that nourish the algae. Scientists have found that corals hosting algae can form reefs up to ten times faster than those without this relationship.

But when coastal waters get unusually hot for long stretches, corals eject the algae in what biologists call a bleaching event. Wide stretches of reefs turn bone white, and if the condition lasts long enough, the corals die. In 2014, a sputtering El Niño warming of the Pacific combined with the general long-term warming of ocean waters from climate change triggered widespread bleaching. But that was just the beginning. The El Niño of 2015–16, which tied as the strongest in modern times with the Pacific warming of 1997–98, sustained and worsened the bleaching in many regions.

In 2017, the National Oceanic and Atmospheric Administration (NOAA) Coral Reef Watch team reported that the added ocean heating from human-caused global warming was almost assuredly contributing to this phenomenon, making this the most widespread and prolonged bleaching on record.

The event had a particularly damaging impact on Australia's Great Barrier Reef, including the reef's northernmost section, which, according to scientists, had never bleached before.

Marine biologists have stressed that corals are remarkably resilient, noting, for example, that deeper portions of reefs remained healthy. But the prolonged episode also drew fresh attention to the importance of cutting other threats to reefs, like runoff from agriculture, untreated sewage, and overfishing. Intensified conservation efforts could help sustain these extraordinarily bountiful and diverse ecosystems through this century and beyond.

SEE ALSO: Lethal Heat and the "Great Dying" (252 million BCE), Global Warming Becomes News (1988)

**A 2016 photograph of coral bleaching around Heron Island**, close to the southernmost point of Australia's Great Barrier Reef. As a three-year global bleaching episode ended in 2017, scientists said it was the widest and worst yet measured.

# AN END TO ICE AGES?

102,018 CE

I N THE 1970S, HEADLINES BRIEFLY FOCUSED on the chilling prospect of a devastating climate change—in an icy direction. Some scientists assessing recent fluctuations in global temperature proposed that, after more than 11,000 years of relative warmth since the end of the last ice age, another big chill could be in the offing. Further study quickly made it clear that the temperature changes were more of a wiggle than the start of a sustained plunge. Additional research then pointed to a likely end to ice ages altogether for perhaps 100,000 years or more, because of the rising human contribution to the greenhouse effect. In a 2000 paper, the Belgian scientists André Berger and Marie-France Loutre calculated that warming from added heat-trapping gases could easily overwhelm the cooling influence of the wobbles in Earth's orbit and rotation linked to past ice age cycles.

More science has since reinforced that whatever debates persist around the pace and extent of human-driven change in this century, and whatever choices societies make about fossil fuels and $CO_2$-free energy alternatives, our influence on the climate already guarantees tens of thousands of years of warmer conditions and rising sea levels (barring some geoengineering advance or an asteroid strike).

In a book published two years after he helped organize maritime weather science in 1853, the oceanographer Matthew Fontaine Maury (1806–73) included a passage that captures what it means to develop a new relationship with climate:

> [T]he atmosphere is something more than a shoreless ocean. . . . It is an inexhaustible magazine, marvelously adapted for many benign and beneficent purposes. Upon the proper working of this machine depends the well being of every plant and animal that inhabits the earth; therefore the management of it, its movements, and the performance of its offices, cannot be left to chance.

The core challenge lies in finding ways to cut the odds of worst-case climate outcomes while still satisfying humanity's energy needs. And so the story will continue—with many more milestones in weather and climate history to come.

SEE ALSO: Orbits and Ice Ages (1912), A President's Climate Warning (1965), Climate by Design? (2006)

**A view of the western flanks of Greenland's vast ice sheet** in 2004. Scientists have calculated that the buildup of long-lasting greenhouse gases from human activities is almost sure to delay the advance of these and other ice sheets in a new ice age for tens of millennia to come.

# Contributors

I**N SHAPING THIS BOOK, WE DECIDED EARLY ON** to invite some friends and scholars with specialized knowledge to contribute certain items. Once we decided to start the chronology with the research illuminating the early moments in the history of Earth's atmosphere and climate, it was logical to hand things over for a bit to **Howard Lee**, who has degrees in geology and remote sensing and is the author of *Your Life as Planet Earth: A New Way to Understand the Story of the Earth, Its Climate and Our Origins* (Amazon Kindle, 2014). Lee, who wrote items on the first 4.533 billion years of Earth history and one more, on the rise of agriculture, is also a fellow of the Geological Society of London. Explore his website for his book: ylape.com.

**Stephen T. Jackson, PhD**, professor emeritus of botany and ecology at the University of Wyoming, wrote the piece in praise of the Danish naturalist Japetus Steenstrup. He is a fellow of the Ecological Society of America and the American Association for the Advancement of Science. Jackson has edited two translations of foundational works by Alexander von Humboldt—*Essay on the Geography of Plants* (University of Chicago Press, 2009) and *Views of Nature* (University of Chicago Press, 2014)—with a third on the way.

**Professor B. Lynn Ingram, PhD**, who wrote the piece on California's great deluge, is an expert on using geological clues to reveal past climate and weather events. She is professor emerita at the University of California, Berkeley, and co-author, with Frances Malamud-Roam, of *The West without Water: What Past Floods, Droughts, and Other Climate Clues Tell Us about Tomorrow* (University of California Press, 2013).

**John Schwartz**, a longtime *New York Times* reporter, author and native of Galveston, Texas, contributed the item on that city's "Mighty Storm." Read his Times stories at nytimes.com/by/john-schwartz.

**Curt Stager, PhD**, contributed the piece on Africa's super drought, drawing on his own influential research. While his fieldwork focuses on the past, Stager, who is a professor of natural sciences at Paul Smith's College in the Adirondacks, also looks ahead, as he did so well in *Deep Future: The Next 100,000 Years of Life on Earth* (St. Martin's/Thomas Dunne, 2011).

**Graeme L. Stephens, PhD**, who wrote about Luke Howard, the man who named the clouds, directs the Center for Climate Sciences at NASA's Jet Propulsion Laboratory in Pasadena, California.

**Paul D. Williams, PhD**, contributed the milestone on Lewis Fry Richardson's precursor to climate models. Williams is Professor of Atmospheric Science and Royal Society University Research Fellow in the Department of Meteorology at the University of Reading.

# References

Here are the main references for the 100 moments and milestones in this book, along with the names of guest contributors where relevant. For a full list, please contact Andrew Revkin by email at revkin+weather@gmail.com.

**4.567 Billion BCE: Earth Gets an Atmosphere**
*Contributed by Howard Lee*
Connelly, James N., Martin Bizzarro, Alexander N. Krot, Åke Nordlund, Daniel Wielandt, and Marina A. Ivanova. "The Absolute Chronology and Thermal Processing of Solids in the Solar Protoplanetary Disk." *Science*, Nov 2012. http://bit.ly/2vI0HGI.
Walsh, Kevin J., and Harold F. Levison. *Terrestrial Planet Formation from an Annulus*. The American Astronomical Society. Sept 2016. http://bit.ly/2vcDpXI.

**4.3 Billion BCE: Water World**
*Contributed by Howard Lee*
Valley, John W., et al. "Hadean age for a post-magma-ocean zircon confirmed by atom-probe tomography." *Nature Geoscience*, Feb 2014. http://go.nature.com/1q92mcM.
Zahnle, Kevin, Norman H. Sleep, et al. "Emergence of a Habitable Planet." *Space Science Reviews*, March 2007.

**2.9 Billion BCE: Pink Skies and Ice**
*Contributed by Howard Lee*
Dell'Amore, Christine, and Robert Kunzig. "Why Ancient Earth Was So Warm." *National Geographic*, July 2013. http://bit.ly/2uzQR4A.

**2.7 Billion BCE: First Fossil Traces of Raindrops**
*Contributed by Howard Lee*
Charnay, B., F. Forget, R. Wordsworth, J. Leconte, E. Millour, F. Codron, and A. Spiga. "Exploring the faint young Sun problem and the possible climates of the Archean Earth with a 3-D GCM." *Journal of Geophysical Research*, 19 Sept 2013 http://bit.ly/2uhMWK8.
Som, Sanjoy M., Roger Buick, James W. Hagadorn, Tim S. Blake, John M. Perreault, Jelte P. Harnmeijer, and David C. Catling. "Earth's air pressure 2.7 billion years ago constrained to less than half of modern levels." *Nature Geoscience*, 09 May 2016. http://go.nature.com/2vFwQO7.

**2.4 Billion–423 Million BCE: The Icy Path to Fire**
*Contributed by Howard Lee*
Brocks, J. J., A. J. M. Jarrett, E. Sirantoine, F. Kenig, M. Moczydłowska, S. Porter, and J. Hope. "Early sponges and toxic protists: possible sources of Cryostane, an age diagnostic biomarker antedating Sturtian Snowball Earth." *Geobiology*, 28 Oct 2015. http://bit.ly/2vxvFRe.
Cole, Devon B., Christopher T. Reinhard, Xiangli Wang, Bleuenn Gueguen, Galen P. Halverson, Timothy Gibson, Malcolm S.W. Hodgskiss, N. Ryan McKenzie, Timothy W. Lyons, and Noah J. Planavsky. "A shale-hosted Cr isotope record of low atmospheric oxygen during the Proterozoic." *Geology*, 17 May 2016. http://bit.ly/2vFUpqf.

**252 Million BCE: Lethal Heat and the Permian "Great Dying"**
*Contributed by Howard Lee*
Bond, David P. G., Paul B. Wignall, Michael M. Joachimski, Yadong Sun, Ivan Savov, Stephen E. Grasby, Benoit Beauchamp, and Dierk P. G. Blomeier. "An abrupt extinction in the Middle Permian (Capitanian) of the Boreal Realm (Spitsbergen) and its link to anoxia and acidification." The Geological Society of America. 4 March 2015. http://bit.ly/1CLrmL4.
Sun, Yadong, Michael M. Joachimski, Paul B. Wignall, Chunbo Yan, Yanlong Chen, Haishui Jiang, Lina Wang, and Xulong Lai. "Lethally Hot Temperatures During the Early Triassic Greenhouse." *Science*, 19 Oct 2012. http://bit.ly/2vImded.

**66 Million BCE: Dinosaurs' Demise, Mammals Rise**
*Contributed by Howard Lee*
*The Cenozoic Era*. University of California Museum of Paleontology. 2008. http://bit.ly/1kLrPpD.
Petersen, Sierra V., Andrea Dutton, and Kyger C. Lohmann. "End-Cretaceous extinction in Antarctica linked to both Deccan volcanism and meteorite impact via climate change." *Nature Communications*. 05 July 2016. http://go.nature.com/2wmRSz3.

**56 Million BCE: The Feverish Eocene**
*Contributed by Howard Lee*
Alley, Richard B. "A heated mirror for future climate." *Science*, 08 Apr 2016. http://bit.ly/1STeLPY.
Harrington, Guy J., and Carlos A. Jaramillo. "Paratropical floral extinction in the Late Palaeocene–Early Eocene." *Journal of the Geological Society*, 23 June 2006. http://bit.ly/2hCMjcx.

**34 Million BCE: A Southern Ocean Chills Things**
*Contributed by Howard Lee*
Lear, Caroline H., and Dan J. Lunt. "How Antarctica got its ice."

*Science*. 01 April 2016. http://bit.ly/2vcKdou.

**10 Million BCE: The Rise of Tibet and the Asian Monsoon**
Hu, Xiumian, Eduardo Garzanti, Jiangang Wang, Wentao Huang, and Alex Webb. "The timing of India-Asia collision onset—Facts, theories, controversies." *ScienceDirect*, 29 July 2016. http://bit.ly/2vcEr6e.
Sun, Youbin, Long Ma, Jan Bloemendal, Steven Clemens, Xiaoke Qiang, and Zhisheng An. "Miocene climate change on the Chinese Loess Plateau: Possible links to the growth of the northern Tibetan Plateau and global cooling." *Geochemistry, Geophysics, Geosystems*, July 2015.

**100,000 BCE: Climate Pulse Propels Populations**
Irfan, Umair. "Climate Change May Have Spurred Human Evolution." *Scientific American*, 02 Jan 2013. http://bit.ly/2wAWeBI.
Maslina, Mark A., Chris M. Brierley, Alice M. Milner, Susanne Shultz, Martin H. Trauth, and Katy E. Wilson. "East African climate pulses and early human evolution." *Quaternary Science Reviews*, 12 June 2016. http://bit.ly/2uzfFcO.
deMenocal, Peter B., and Chris Stringer. "Human migration: Climate and the peopling of the world." *Nature*, 21 Sept 2016. http://go.nature.com/2dSJnCM.

**15,000 BCE: A Super Drought**
*Contributed by Curt Stager*
Stager, J. Curt, David B. Ryves, Brian M. Chase, and Francesco S. R. Pausata. "Catastrophic Drought in the Afro-Asian Monsoon Region During Heinrich Event 1." *Science*, 24 Feb 2011. http://bit.ly/2uz4Mrq.

**9,700 BCE: The Fertile Crescent**
"Did Climate Change Help Spark The Syrian War? Scientists Link Warming Trend to Record Drought and Later Unrest." The Earth Institute, Columbia University. 02 March 2015. http://bit.ly/2vcZkOQ.
Sharifi, Arash, Peter K. Swart et al. "Abrupt climate variability since the last deglaciation based on a high-resolution, multi-proxy peat record from NW Iran: The hand that rocked the Cradle of Civilization?" *Quaternary Science Reviews*, 01 Sept 2015. http://bit.ly/2vHNYnc.

**5,300 BCE: North Africa Dries and the Pharaohs Rise**
Allen, Susie, and William Harms. "World's oldest weather report could revise Bronze Age chronology." *UChicago News*, 01 April 2014. http://bit.ly/2hDfSKU.
deMenocal, Peter B., and J. E. Tierney. "Green Sahara: African Humid Periods Paced by Earth's Orbital Changes." *Nature Education Knowledge*. http://bit.ly/2vxsKb9.

**5,000 BCE: Agriculture Warms the Climate**
*Contributed by Howard Lee*
Ruddiman, W. F., et al. "Late Holocene climate: Natural or anthropogenic?" *Reviews of Geophysics*, March 2016. http://bit.ly/2vFMTeS.
Stanley, Sarah. "Early Agriculture Has Kept Earth Warm for Millennia." *Review of Geophysics*, 19 Jan 2016. http://bit.ly/2vxtsp7.

**350 BCE: Aristotle's *Meteorologica***
Aristotle. *Meteorologica*. Translated by E. W. Webster. Internet Classic Archives. http://www.ucmp.berkeley.edu/history/aristotle.html.

**300 BCE: China Shifts from Mythology to Meteorology**
Doggett, L. E. *Calendars*. NASA Goddard Space Flight Center.
Needham, Joseph. *Science and Civilization in China*. United Kingdom: Cambridge University Press, 2000.

**1088 CE: Shen Kuo Writes of Climate Change**
Edwards, Steven A., Ph.D. "Shen Kuo, the first Renaissance man?" American Association for the Advancement of Science. 15 March 2012. http://bit.ly/2uhgYOw.
"The first evidence for climate change." Geological Society of London blog. 03 March 2014. http://bit.ly/2fmjsIE

**1100: Medieval Warmth to a Little Ice Age**
Bradley, Raymond S., Heinz Wanner, and Henry F. Diaz. "The Medieval Quiet Period." *The Holocene*, 22 Jan 2016. http://bit.ly/2wmJC23.
Camenisch, Chantal, et al. "The 1430s: a cold period of extraordinary internal climate variability during the early Spörer Minimum with social and economic impacts in north-western and central Europe." *Climate of the Past*, 01 Dec 2016. http://bit.ly/2hgnYZx.

**1571: The Age of Sail**
"Naval history of China." Wikipedia. Accessed June 5, 2017. http://bit.ly/2uzhDdc.
"Winds of Change: Defeat of the Spanish Armada, 1588." *NASA Landsat Science*, 25 May 2017. https://go.nasa.gov/2uyVa03.

**1603: The Invention of Temperature**
Van Helden, Al. "The Thermometer." The Galileo Project. Rice University. 1995. http://bit.ly/2tVRRQa.

Williams, Matt. "What Did Galileo Invent?" *Universe Today*. Nov. 21, 2016. http://bit.ly/2fm7f6F.

**1637: Deciphering the Rainbow**
Butterworth, Jon. "How the rainbow illuminates the enduring mystery of physics." *Aeon*. Jan. 3, 2017. http://bit.ly/2vFVxKd.
Haußmann, Alexander. "Rainbows in nature: recent advances in observation and theory." *European Journal of Physics* 37, no. 6 (August 26, 2016). http://bit.ly/2fcCjRc.

**1644: The Weight of the Atmosphere**
O'Connor, J. J., and E. F. Robertson. "Evangelista Torricelli." Nov 2002. http://bit.ly/2vHSUIO.
Williams, Richard. "October, 1644: Torricelli Demonstrates the Existence of a Vacuum Elegant physics experiment; enduring practical invention." *American Physical Society*, Oct 2012. http://bit.ly/2uhAGJH.

**1645: A Spotless Sun**
Degroot, Dagomar. "What was the Maunder Minimum? New Perspectives on an Old Question." Historicalclimatology.com. 9 June, 2016. http://bit.ly/1XSaxNK.
Meehl, Gerald A., Julie M. Arblaster, and Daniel R. Marsh. "Could a future 'Grand Solar Minimum' like the Maunder Minimum stop global warming?" *Geophysical Research Letters: An AGU Journal*, 13 May 2013. http://bit.ly/2hDAtPs.

**1714: Fahrenheit Standardizes Degrees**
Chang, Hasok. *Inventing Temperature: Measurement and Scientific Progress*. New York and Oxford: Oxford University Press, 2004.

Radford, Tim. "A Brief History of Thermometers." *The Guardian*, 06 Aug 2003. http://bit.ly/2hCODQT.
"Temperature and Temperature Scales." World of Earth Science. Encyclopedia.com. 04 Jun. 2017. http://www.encyclopedia.com/science/encyclopedias-almanacs- transcripts-and-maps/temperature-and-temperature-scales.

**1721: Four Seasons on Four Strings**
Ortiz, Edward. "Taking the World By Storm? Weather Inspired Music." *San Francisco Classical Voice*, 30 July 2013. http://bit.ly/1k6vH2T.
St. George, Scott, Daniel Crawford, Todd Reubold, and Elizabeth Giorgi. "Making Climate Data Sing: Using Music-like Sonifications to Convey a Key Climate Record." *American Meteorological Society*. Jan 2017. http://bit.ly/2uikPip.

**1735: Mapping the Winds**
"Meteorology/Edmond Halley, 1656–1742." Princeton University. http://bit.ly/1xcs6cZ.
Persson, Anders O. "Hadley's Principle: Understanding and Misunderstanding the Trade Winds." 2006.

**1752: Benjamin Franklin's Lightning Rod**
"Franklin's Lightning Rod." The Franklin Institute. 2017. http://bit.ly/1TFjJlr.
"The Kite Experiment, 19 October 1752." Founders Online, National Archives, last modified 30 March 2017. http://bit.ly/2s59f7a.
Krider, E. Philip. "Benjamin Franklin and the First Lightning Conductors." *Proceedings of the International Commission on History of Meteorology*. Volume 1 (2004).

**1755: Franklin Chases a Whirlwind**
Heidorn, Keith C., Ph.D. "Benjamin Franklin: The First American Storm Chaser." *The Weather Doctor*, 1998. http://bit.ly/2wALqDT.

**1783: First Weather Balloon Flight**
"A Brief History of Upper Air Observations." NOAA National Weather Service. http://bit.ly/2wAWPUj.
*Léon Teisserenc de Bort*. Encyclopedia Brittanica, Inc. http://bit.ly/2fmfPlZ.

**1792: *The Farmer's Almanac***
Hale, Justin. "Predicting Snow for the Summer of 1816, The Year Without a Summer." *The Old Farmer's Almanac*. http://bit.ly/2vGjFfO.
"History of the Old Farmer's Almanac. The Almanac Editors' Legacies." *The Old Farmer's Almanac*. http://bit.ly/2hBH8cG.

**1802: Luke Howard Names the Clouds**
*Contributed by Graeme L. Stephens*
Howard, Luke. *Essay on the Modifications of Clouds*. London: Churchill & Sons, 1803. http://bit.ly/1IAJdpS.
Stephens, Graeme L. "The Useful Pursuit of Shadows." *American Scientist*, Sept 2003.

**1802: Humboldt Maps a Connected Planet**
Von Humboldt, Alexander, and Aimé Bonpland. *Essay on the Geography of Plants*, edited by Stephen L Jackson. Translated by Sylvie Romanowski. Chicago: University of Chicago Press, 2009.
Wulf, Andrea. *The Invention of Nature: Alexander von Humboldt's New World*. New York: Alfred A. Knopf, 2015.

**1806: Beaufort Classifies the Winds**
"Beaufort Wind Scale." NOAA. https://www.weather.gov/mfl/beaufort.
"Wind Measurements." Weather for Schools. http://bit.ly/2vIppGJ.

**1814: London's Last Frost Fair**
de Castella, Tom. "Frost fair: When an elephant walked on the frozen River Thames." *BBC News Magazine*, 8 Jan 2014. http://bbc.in/2cxo4o2.
Johnson, Ben. "The Thames Frost Fairs." *Historic U.K.*, 2017. http://bit.ly/1CMxgyj.

**1816: An Eruption, Famine, and Monsters**
Buzwell, Greg. "Mary Shelley, *Frankenstein* and the Villa Diodati." *British Library*, 15 May 2014. http://bit.ly/2myuQk0.
Cavendish, Richard. "The eruption of Mount Tambora." *History Today*, April 2015. http://bit.ly/1r7PfgO.
Evans, Robert. "The eruption of Mount Tambora killed thousands, plunged much of the world into a frightful chill and offers lessons for today." *Smithsonian*, July 2002. http://bit.ly/1sPH2gw.

**1818: Watermelon Snow**
Edwards, Howell G.M., Luiz F.C. de Oliveira, Charles S. Cockell, J. Cynan Ellis-Evans, and David D. Wynn-Williams. "Raman spectroscopy of senescing snow algae: pigmentation changes in an Antarctic cold desert extremophile." *International Journal of Astrobiology*, April 2004.
Frazer, Jennifer. "Wonderful Things: Don't Eat the Pink Snow." *Scientific American*, 9 July 2013. http://bit.ly/2hBHJuW.

**1830: An Umbrella for Everyone**
"Samuel Fox, Bradwell's Most Famous Son." The Samuel Fox Country Inn. http://bit.ly/2vcH8ov.
Sangster, William. *Umbrellas and their History*. London: Oxford University Press. Published online by Project Gutenberg. http://bit.ly/2uicET0.

**1840: Ice Ages Revealed**
Bressan, David. "The discovery of the ruins of ice: The birth of glacier research." *Scientific American*, 03 Jan 2011. http://bit.ly/2gFkR87.
Reebeek, Holli. "Paleoclimatology: Introduction." NASA Earth Observatory. 28 June 2005. https://go.nasa.gov/2uiA0YE.
Summerhayes, C. P. *Earth's Climate Evolution*. Wiley Blackwell, 2015. http://bit.ly/1N5QQId.

**1841: Peat Bog History**
*Contributed by Stephen Jackson*
Jackson, Stephen T., and Dan Charman. "Editorial: Peatlands—paleoenvironments and carbon dynamics." *PAGES News*, April 2010. http://bit.ly/2uzpFmo.

**1845: Cold Dooms an Arctic Explorer**
Revkin, Andrew C. "Where Ice Once Crushed Ships, Open Water Beckons." *The New York Times*, 24 Sept 2016. http://nyti.ms/2fmrhOD.
"Study offers new insights to the Franklin Expedition mystery." University of Glasgow. *University News*, 22 Sept 2016. http://bit.ly/2vFZNcX.
Watson, Paul. *Ice Ghosts: The Epic Hunt for the Lost Franklin Expedition*. New York: W. W. Norton & Company, 2017.

**1856: Scientists Discover Greenhouse Gases**
Darby, Megan. "Meet the woman who first identified the greenhouse effect." *Climate Home*, 09 Feb 2016. http://bit.ly/2c5Bqsb.
Wogan, David. "Why we know about the greenhouse gas effect." *Scientific American*, 16 May 2013. http://bit.ly/2veIuxb.

**1859: Space Weather Comes to Earth**
Moore, Nicole Casal. "Solar storms: Regional forecasts set to begin." University of Michigan. *Michigan News*, 28 Sept 2016. http://bit.ly/2uiIxuQ.
Orwig, Jessica. "The White House is prepping for a single weather event that could cost $2 trillion in damage." *Business Insider*, 06 Nov 2015. http://read.bi/1NhjPs6.
Philips, Dr. Tony. "Near Miss: The Solar Superstorm of July 2012." NASA. 23 July 2014. https://go.nasa.gov/2wmF7EJ.

**1861: First Weather Forecasts**
Corfidi, Stephen. "A Brief History of the Storm Prediction Center." NOAA. 12 Feb 2010. http://www.spc.noaa.gov/history/early.html.
"Robert FitzRoy and the Daily Weather Reports." Met Office. Last updated 4 May 2016. http://bit.ly/2hCu3zT.

**1862: California's Great Deluge**
*Contributed by B. Lynn Ingram*
Ingram, B. Lynn. "California Megaflood: Lessons from a Forgotten Catastrophe." *Scientific American*, 01 Jan 2013. http://bit.ly/2jk7lMI.

**1870: Meteorology Gets Useful**
"History of The National Weather Service." NOAA National Weather Service. http://www.weather.gov/timeline.
"WFO Juneau's and National Weather Service History." NOAA National Weather Service. weather.gov/ajk/OurOffice-History.

**1871: Midwestern Firestorms**
"1871 Massive fire burns in Wisconsin." *History*, 2009. http://bit.ly/2veA0WX.
Pernin, Peter. "The great Peshtigo fire: an eyewitness account." Madison: State Historical Society of Wisconsin, 1971. http://bit.ly/2vcBnXU.
"The Peshtigo Fire." NOAA National Weather Service. https://www.weather.gov/grb/peshtigofire.

**1880: "Snowflake" Bentley**
Blanchard, Duncan C. "The Snowflake Man." *Weatherwise*, 1970. http://snowflakebentley.com/sfman.htm.
"Wilson A. Bentley: Pioneering Photographer of Snowflakes." Smithsonian Institution Archives. http://s.si.edu/2vG0Cm3.

**1882: Coordinating Arctic Science**
"History of the Previous Polar Years." *Internationales Polarjahr*, 2006. http://bit.ly/2uzHL7N.
Revkin, Andrew. *The North Pole Was Here: Puzzles and Perils at the Top of the World*. Boston: Kingfisher, 2006.

**1884: First Photographs of Tornadoes**
Potter, Sean. "Retrospect: April 26, 1884: Earliest Known Tornado Photograph." *WeatherWise*, March–April 2010. http://bit.ly/2uhFzTi.
Snow, John T. "Early Tornado Photographs." *Journal of the American Meteorological Society*, April 1984. http://bit.ly/2vFYkU0.

**1886: Groundhog Day**
"About Groundhog Day." Website of the Punxsutawney Groundhog Club. http://bit.ly/1tsVlsf.
"Groundhog Day." NOAA National Centers for Environmental Information. http://bit.ly/1KgpLX2.

Wordle, Lisa. "How often is Punxsutawney Phil right? Analysis of Groundhog Day predictions since 1898." PennLive website. 30 Jan 2017. http://bit.ly/2k2VSRg.

**1887: Putting Wind to Work**
"History of Wind Energy." Wind Energy Foundation. 2016.
Price, Trevor J. "Britain's First Modern Wind Power Pioneer." *Wind Engineering Journal*, 01 May 2005.

**1888: The Great White Hurricane**
"Major Winter Storms." NOAA National Weather Service. http://www.weather.gov/aly/MajorWinterStorms.

**1888: Deadliest Hailstorm**
"Highest Mortality Due to Hail." World Meteorological Organization's World Weather & Climate Extremes Archive. 2017. http://bit.ly/2vI7bp0.
"Roopkund Lake's skeleton mystery solved! Scientists reveal bones belong to 9th century people who died during heavy hail storm." *India Today*, 31 May 2013. http://bit.ly/2vxcaZ0.

**1896: First International Cloud Atlas**
"International Cloud Atlas Manual on the Observation of Clouds and Other Meteors." World Meteorological Organization website. https://www.wmocloudatlas.org/.
MacLellan, Lila. "Amateur cloud-spotters lobbied to add this beautiful new cloud to the International Cloud Atlas." *Quartz*, March 2017. http://bit.ly/2mCus7A.

**1896: Coal, CO$_2$, and the Climate**
Fleming, James Rodgers. *Historical Perspectives on Climate Change*. New York: Oxford University Press, 1998.
Weart, Spencer R. *The Discovery of Global Warming*. Boston: Harvard University Press, 2008. http://history.aip.org/climate/.

**1900: A Mighty Storm**
*The 1900 Storm*. Published in conjunction with the City of Galveston 1900 Storm Committee. 2014 Galveston Newspapers Inc. All rights reserved. http://www.1900storm.com/.

**1902: "Manufactured Weather"**
Buchanan, Matt. "An Apparatus for Treating Air: The Modern Air Conditioner." *The New Yorker*, 05 June 2013. http://bit.ly/2vcKqYy.
"History of Air Conditioning." Department of Energy. 20 July 2015. https://energy.gov/articles/history-air-conditioning.
"The Invention that Changed the World." Willis Carrier website. http://www.williscarrier.com/1876–1902.php.

**1903: The Windshield Wiper**
Anderson, Mary. *U.S. Patent No. 743,801: Window-cleaning device*. U.S. Patent and Trademark Office. 18 June 1903. https://www.google.com/patents/US743801.
Slater, Dashka. "Who Made That Windshield Wiper?" *The New York Times Magazine*, 12 Sept 2014.

**1903: A Dry Discovery**
Bortman, Henry. "A Tale of Two Deserts." *Astrobiology Magazine*, 18 April 2011. http://bit.ly/2uzjlvk.
Khan, Alia. "Exploring the Dry Valleys, Then and Now." *The New York Times*, 21 Dec 2011. http://nyti.ms/2uhjhkw.

**1911: The Great Blue Norther**
Samenow, Jason. "Wild rides: the 11/11/11 Great Blue Norther and the largest wave ever surfed." *The Washington Post*, 11 Nov 2011.
University Of Missouri staff. "MU Scientists Detail Cause of 1911 Storm." KOMU. 07 Nov 2011. http://bit.ly/2wn4LZY.

**1912: Orbits and Ice Ages**
Croll, James. *Climate and time in their geological relations; a theory of secular changes of the earth's climate*. New York: D. Appleton, 1875. http://bit.ly/2uzamKu.
Weart, Spencer R. "Past Climate Cycles: Ice Age Speculations." *The Discovery of Global Warming*, updated Jan 2017. https://history.aip.org/climate/cycles.htm.

**1922: A "Forecast Factory"**
*Contributed by Paul D. Williams*
Lynch, Peter. "The origins of computer weather prediction and climate modeling." *Journal of Computational Physics*, 20 March 2008.
Richardson, Lewis Fry. *Weather prediction by numerical process*. Cambridge, MA: Cambridge University Press, 1922.

**1931: "China's Sorrow"**
Chen, Yunzhen, James P. M. Syvitski, Shu Gao, Irina Overeem, and Albert J. Kettner. "Socio-economic Impacts on Flooding: A 4000-Year History of the Yellow River." *China Ambio*, Nov 2012. http://bit.ly/2icuxgl.
Hudec, Kate. "Dealing with the Deluge." NOVA, 26 March 1996. http://to.pbs.org/2vG9E2d.
Wang, Shuai, Bojie Fu, Shilong Piao, Yihe Lü, Philippe Ciais, Xiaoming Feng, and Yafeng Wang. "Reduced sediment transport in the Yellow River due to anthropogenic changes." *Nature Geoscience*, 30 Nov 2015. http://go.nature.com/2flWtgY.

**1934: The Fastest Wind Gust**
"World: Maximum Surface Wind Gust." World Meteorological Organization's World Weather & Climate Extremes Archive. http://bit.ly/2vIdwRg.
"World Record Wind." Mt. Washington Observatory. http://bit.ly/1AbMVoC.

**1935: The Dust Bowl**
"The Black Sunday Dust Storm of April 14, 1935." NOAA National Weather Service. http://bit.ly/2vHKKAl.
Cook, Ben, Ron Miller, and Richard Seager. "Did dust storms make the Dust Bowl drought worse?" The Trustees of Columbia University in the City of New York, Lamont-Doherty Earth Observatory. 2011.

**1941: Russia's "General Winter"**
*Hitler's Table Talk, 1941–1944: His Private Conversations.* Translated by Norman Cameron and R. H. Stevens. New York: Enigma Books, 2008.
Roberts, Andrew. *The Storm of War*. United Kingdom: Telegraph Books. 6 Aug 2009.

**1943: Hurricane Hunters**
Fincher, Lew, and Bill Read. "The 1943 surprise hurricane." NOAA History. April 2017. http://bit.ly/1D48iOc.
"Frequently asked Questions. " The Hurricane Hunters Association. http://www.hurricanehunters.com/faq.htm.
"The Lost Hurricane/Typhoon Hunters: In Memoriam." *Weather Wunderground*, April 2017. http://bit.ly/2uhJwrl.

**1944: The Jet Stream Becomes a Weapon**
Hornyak, Tim. "Winds of war: Japan's balloon bombs took the Pacific battle to American soil." *Japan Times*, 25 July 2015.

Lewis, John M. "Ōishi's Observation Viewed in the Context of Jet Stream Discovery." *Bulletin of the American Meteorological Society*. March 2003. http://bit.ly/2uihfEW.

**1946: Rainmakers**
Fleming, James Rodger. *Fixing the Sky: The Checkered History of Weather and Climate Control*. New York: Columbia University Press, 2010.
Moseman, Andrew. "Does cloud seeding work? China takes credit for the storms now bringing a reprieve from severe drought, but is that claim valid?" *Scientific American*, 19 Feb 2009. http://bit.ly/2uiOfNc.

**1950: The First Computerized Forecast**
"Electronic Computer Project." Institute for Advanced Study. 2017. https://www.ias.edu/electronic-computer-project.
Lynch, Peter. "The Origins of Computer weather Prediction and Climate Modeling." *Journal Of Computational Physics*, 19 March 2007. http://bit.ly/2vI8zba.

**1950: Tornado Warnings Advance**
Angel, Jim. "60th Anniversary of the First Tornado Detected by Radar." *Illinois State Climatologist*, 09 April 2013. http://bit.ly/2wmDBCp.
Smith, Mike. *Warnings: The True Story of How Science Tamed the Weather*. Greenleaf Book Group, 2010.

**1952: London's Great Smog**
"The Great Smog of 1952." Met Office web page. 20 April 2015. http://bit.ly/1OVCbTO.

**1953: North Sea Flood**
Weesjes, Elke. "The 1953 North Sea Flood in the Netherlands, Impact and Aftermath." *Natural Hazards Observer*, 28 Sept 2015. http://bit.ly/2vGlQzU.

**1958: The Rising Curve of $CO_2$**
"The Keeling Curve." Scripps Institution of Oceanography. https://scripps.ucsd.edu/programs/keelingcurve/.
Weart, Spencer. *The Discovery of Global Warming*. Boston: Harvard University Press, 2008.

**1960: Watching Weather from Orbit**
Alfred, Randy. "April 1, 1960: First Weather Satellite Launched." *Science*, 04 Jan 2008. http://bit.ly/2fmMwQr.
"NOAA's GOES-16 satellite sends first images of Earth; Higher-resolution details will lead to more accurate forecasts." NOAA. 23 Jan 2017. http://bit.ly/2jQhzUD.

**1960: Chaos and Climate**
Dizikes, Peter. "When the Butterfly Effect Took Flight." *MIT Technology Review*, 22 Feb 2011. http://bit.ly/2tOmXdM.
Fleming, James Rodger. *Inventing Atmospheric Science*. Cambridge, MA: MIT Press, 2016.
Gleick, James. *Chaos: Making a New Science*. New York: Viking, 1987.
Lorenz, Edward N., Sc.D. "Does the Flap of a Butterfly's wings in Brazil set off a Tornado in Texas?" American Association for the Advancement of Science, 139th meeting. 29 March 1972. http://bit.ly/1eUrMno.

**1965: A President's Climate Warning**
Johnson, Lyndon Baines. "Special Message to Congress on Conservation and Restoration of Natural Beauty." 08 Feb 1965. http://bit.ly/2k3XrzX.

Lavelle, Marianne. "A 50th anniversary few remember: LBJ's warning on carbon dioxide." *The Daily Climate,* 02 Feb 2015. http://bit.ly/1uQSeFX.

**1967: Climate Models Come of Age**
Pidcock, Roz. "The most influential climate change papers of all time." *Carbon Brief,* 06 July 2015. http://bit.ly/2qnb8wG.
"Validating Climate Models." National Academy of Sciences. 2012. http://bit.ly/2vxq3Gs.

**1973: Storm Chasing Gets Scientific**
Czuchnicki, Cammie. "History of Storm Chasing." Royal Meteorological Society. http://bit.ly/2wB2XMf.
Golden, Joseph H., and Daniel Purcell. "Life Cycle of the Union City, Oklahoma Tornado and Comparison with Waterspouts." *Monthly Weather Review,* Jan 1978.

**1975: Dangerous Downbursts Revealed**
Fujita, T. Theodore. "Tornadoes and Downbursts in the Context of Generalized Planetary Scales." *Journal of Atmospheric Sciences,* Aug 1981. http://bit.ly/2wmZSjs.
Henson, Bob. "Tornadoes, Microbursts, and Silver Linings: How the Jumbo Outbreak of 1974 helped lead to safer air travel." *AtmosNews,* 01 April 2014. http://bit.ly/2wAYg53.

**1978: Sea Level Threat in Antarctic Ice**
Mercer, J. H. "West Antarctic ice sheet and $CO_2$ greenhouse effect: a threat of disaster." *Nature,* 26 Jan 1978. http://go.nature.com/2fmNonY.

**1983: The Coldest Place on Earth**
Woo, Marcus. "New Record for Coldest Place on Earth, in Antarctica." *National Geographic Magazine,* 11 Dec 2013. http://bit.ly/2wmRRv3.

**1983: Nuclear Winter**
Revkin, Andrew. "Hard Facts About Nuclear Winter." *Science Digest,* March 1985. j.mp/nuclearwinter85.
Turco, R. P., O. B. Toon, T. P. Ackerman, J. B. Pollack, and Carl Sagan. "Nuclear Winter: Global Consequences of Multiple Nuclear Explosions." *Science,* 23 Dec 1983. http://bit.ly/2vIiddS.

**1986: Forecasting El Niño**
Krajick, Kevin. "Mark Cane, George Philander, Win 2017 Vetlesen Prize." Center for Climate and Life, Columbia University. 26 Jan 2017.
McPhaden, Michael. "Predicting El Niño Then and Now." NOAA. 03 April 2015. http://bit.ly/2hBXulA.

**1988: Global Warming Becomes News**
Revkin, Andrew. "Endless Summer: Living with the Greenhouse Effect." *Discover,* Oct 1988. j.mp/greenhouse88.

**1989: Proof of Electric "Sprites"**
Fecht, Sarah. "What Is a Red Sprite? Ghost? Alien? Carbonated beverage?" *Popular Science,* 25 Aug 2015. http://www.popsci.com/what-red-sprite.
Rozell, Ned. "Alaska scientist leaves colorful legacy." University of Alaska, Fairbanks, Geophysical Institute. 01 Feb 2012. http://bit.ly/2uzidYE.

**1993: Climate Clues in Ice and Mud**
Alley, Richard B. *The Two-Mile Time Machine: Ice Cores, Abrupt Climate Change, and Our Future*. Princeton, NJ: Princeton University Press, updated 2014.
Riebeek, Holli. "Paleoclimatology: The Ice Core Revealed." NASA Earth Observatory. 19 Dec 2005. https://go.nasa.gov/2hDO758.

**2006: The Human Factor in Weather Disasters**
Emanuel, Kerry, et al. "Statement on the U.S. Hurricane Problem." Website of Professor Kerry Emanuel, MIT. 25 July 2006. http://bit.ly/2uzDq4r.
Revkin, Andrew C. "Climate Experts Warn of More Coastal Building." *The New York Times*, 25 July 2006.

**2006: Climate by Design?**
Fleming, James Rodger. *Fixing the Sky: The Checkered History of Weather and Climate Control*. New York: Columbia University Press, 2010.
Temple, James. "Harvard Scientists Moving Ahead on Plans for Atmospheric Geoengineering Experiments." *MIT Technology Review*, 24 March 2017. http://bit.ly/2wAXkxJ.

**2006: Long-Distance Dust**
Dunion, Jason P., and Christopher S. Velden. "The Impact of the Saharan Air Layer on Atlantic Tropical Cyclone Activity." American Meteorological Society. *Journals Online*, 01 March 2004. http://bit.ly/2veKF3S.
Kaplan, Sarah. "How dust from the Sahara fuels poisonous bacteria blooms in the Caribbean." *The Washington Post*, 11 May 2016. http://wapo.st/2fnsxAP.

**2007: Tracking the Oceans' Climate Role**
The Argo Project. University of California, San Diego. http://www.argo.ucsd.edu/.
*Evolution of Physical Oceanography*. Edited by Bruce A. Warren and Carl Wunsch. Cambridge, MA: MIT Press, 1981.

**2012: Science Probes the Political Climate**
Kahan, Dan M., Ellen Peters, Maggie Wittlin, Paul Slovic, Lisa Larrimore Ouellette, Donald Braman, and Gregory Mandel. "The polarizing impact of science literacy and numeracy on perceived climate change risks." *Nature Climate Change*, 27 May 2012.

**2013: Settling a Hot Debate**
El Fadli, Khalid I., Randall S. Cerveny, Christopher C. Burt, Philip Eden, David Parker, Manola Brunet, Thomas C. Peterson, Gianpaolo Mordacchini, Vinicio Pelino, Pierre Bessemoulin, José Luis Stella, Fatima Driouech, M. M Abdel Wahab, and Matthew B. Pace. "World Meteorological Organization Assessment of the Purported World Record 58°C Temperature Extreme at El Azizia, Libya (13 September 1922)." American Meteorological Society. Feb 2013.
Samenow, Jason. "Two Middle East locations hit 129 degrees, hottest ever in Eastern Hemisphere, maybe the world." *The Washington Post*, 22 July 2016. http://wapo.st/2wBbNJE.

**2014: The Polar Vortex**
Kennedy, Caitlyn. "Wobbly polar vortex triggers extreme cold air outbreak." NOAA. 8 Jan 2014. http://bit.ly/2vHXff3.
Wiltgen, Nick. "Deep Freeze Recap: Coldest Temperatures of the Century for Some." Weather.com. Accessed 10 Jan 2014. http://wxch.nl/2wmXbyr.

**2015: Climate Diplomacy from Rio through Paris**
Revkin, Andrew C. "The Climate Path Ahead." *The New York Times, Sunday Review*, 12 Dec 2015. http://nyti.ms/2vxNmQJ.

**2016: Arctic Sea Ice Retreat**
Fetterer, Florence. "Piecing together the Arctic's sea ice history back to 1850." *Carbon Brief*, 11 Aug 2016. http://bit.ly/2byE5fC.

**2016: Extreme Lightning**
Lang, Timothy J., Stéphane Pédeboy, William Rison, Randall S. Cerveny, Joan Montanyà, Serge Chauzy, Donald R. MacGorman, Ronald L. Holle, Eldo E. Ávila, Yijun Zhang, Gregory Carbin, Edward R. Mansell, Yuriy Kuleshov, Thomas C. Peterson, Manola Brunet, Fatima Driouech, and Daniel S. Krahenbuhl. "WMO World Record Lightning Extremes: Longest Reported Flash Distance and Longest Reported Flash Duration." American Meteorological Society. June 2017. http://bit.ly/2cPQQpz.

**2017: Reefs Feel the Heat**
"Climate Change Threatens the Survival of Coral Reefs." ISRS Consensus Statement on Climate Change and Coral Bleaching. Oct 2015. http://bit.ly/1MGCoWh.
NOA Satellite and Information Service. Coral Reef Watch. https://coralreefwatch.noaa.gov.

**102,018 CE: An End to Ice Ages?**
Archer, David. *The Long Thaw: How Humans Are Changing the Next 100,000 Years of Earth's Climate*. Princeton, NJ: Princeton University Press, 2008.
Loutre, M. F., and A. Berger. "Future climatic changes: are we entering an exceptionally long interglacial?" *Climatic Change*, July 2000.

# Image Credits

# Index

# About the Authors

**ANDREW REVKIN** is the senior reporter for climate and related issues at the Pulitzer Prize-winning nonprofit newsroom ProPublica. He has written on science and the environment for more than three decades, from the North Pole to the Equator, mainly for the *New York Times,* and won most of the top awards in science journalism, along with a Guggenheim Fellowship. Revkin has written acclaimed books on global warming, the changing Arctic, and the violent assault on the Amazon rain forest. Drawing on his experience with his *Times* blog, Dot Earth, which *Time* magazine named one of the top 25 blogs in 2013, Revkin has spoken to audiences around the world about paths to progress on a turbulent planet. As a young man, he sailed two thirds of the way around the world, encountering waterspouts and potent gales. In spare moments, he is a performing songwriter and was a frequent accompanist for Pete Seeger. Two films have been based on his work: *Rock Star* (Warner Brothers, 2001) and *The Burning Season* (HBO, 1994), which starred Raul Julia and won two Emmy Awards and three Golden Globes.

**LISA MECHALEY** is an educator at the Children's Environmental Literacy Foundation and was formerly the education director at the Hudson Highlands Nature Museum and a middle school science teacher. She and Andrew Revkin have been married for twenty-one years. This is their first book together.

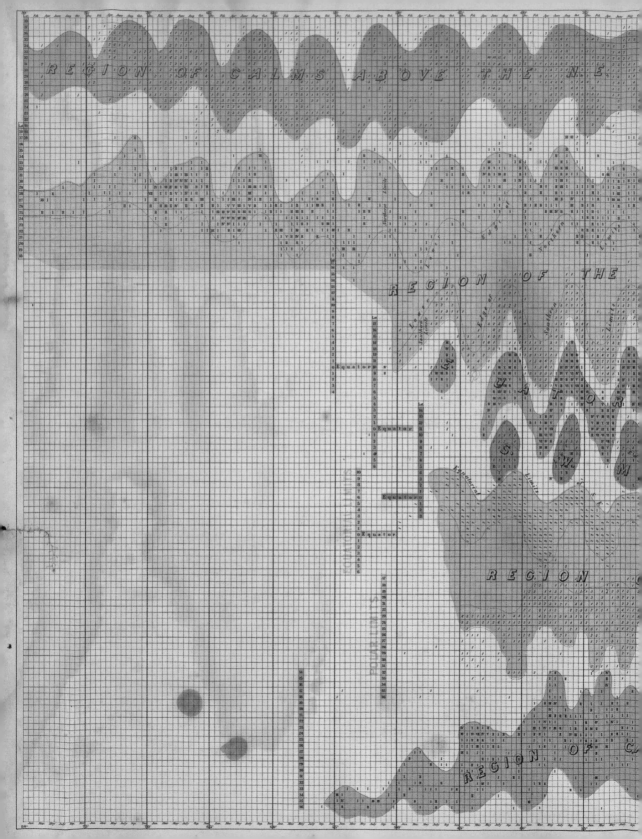

REGION OF CALMS ABOVE THE NE

REGION OF THE

REGION OF C